THE VOYAGES OF
THE 'MORNING'

by
GERALD S. DOORLY

INTRODUCTION BY D. W. H. WALTON

BLUNTISHAM BOOKS · ERSKINE PRESS
1995

First published by Smith, Elder & Co. in 1916
This edition published in 1995 by
Bluntisham Books and The Erskine Press
Banham, Norwich, Norfolk

© Introduction by D. W. H. Walton 1995

ISBN 1 85297 040 5

Printed in Great Britain by
Antony Rowe Limited
Chippenham

INTRODUCTION

Gerald Doorly's chief claim to fame is this account of the two voyages made by the *Morning* to relieve Scott's expedition. There is little point in reiterating the story so well told in this narrative but I shall try to provide some biographical information to introduce the author, although the record is fairly sparse.

Doorly, born in Trinidad on 4 June 1880, was the son of a clergyman. He was brought up in Trinidad, and first came to England when he was nine years old for a six-month holiday with his family. By then he had already decided that the sea was where his interests lay. Taken to South Africa by his father, at the age of fourteen he ran away to sea, sailing to London on a tramp steamer, the *Arno*, to join HMS *Worcester* as a cadet. He was fortunate to be paired up with E R G R (Teddy) Evans, which began a life long friendship. He clearly enjoyed his time as a cadet and at the end of his time was voted the Queen's Gold Medallist by his fellow cadets.

Graduating from *Worcester* Doorly went off as an apprentice on the barque *Auldgirth*, carrying coal from Cardiff to Antofagasta in Chile. Later on the *Auldgirth* was running wheat from California to Port Elizabeth, coal from Australia to Chile and nitrates from Chile to California. Doorly began studying for his Second Mates certificate and, having passed, was appointed as a junior officer in a P & O passenger ship being used as a floating hospital for troops in the South African war. On his return to England in June 1920 he found a telegram from Teddy Evans inviting him to join the *Morning* as Third Officer. Thus was born his participation in the Antarctic voyages of the *Morning*.

After the conclusion of the expedition Doorly took his Master's certificate and joined a banana boat running from Avonmouth to

Jamaica. This lasted only a short time and in 1905 Doorly had resigned and applied to emigrate to New Zealand, where he spent most of the rest of his life. This was clearly connected to his marriage to a New Zealand woman. In New Zealand he became Captain, first of a coaster and later of larger passenger vessels.

Doorly had always been a member of the Royal Naval Reserve and in 1914 joined a troopship leaving for the UK. Soon he was appointed captain of the freighter *Aparima*, leaving in November 1916 with a cargo of high explosives for East Africa. After a succession of successful voyages the *Aparima* was torpedoed in November 1917 off Portland Bill, Dorset, and fifty-four men were lost. He survived unscathed and shortly after returned to New Zealand where he rejoined his original company. In November 1918 he found himself and his crew ill with plague at Tahiti but again he survived.

He maintained his interest in Antarctic matters, giving many lectures in Australia and New Zealand and maintaining a desultory correspondence with his Antarctic shipmates for decades. He died on 3 November 1956.

Music has not been an important feature of Antarctic expeditions but in this the *Morning* was different. The ship had a piano, a gift from Sir Clements Markham. To get it into the *Morning* proved a real problem which was solved by sawing it in half. It survived the trauma and went on to provide a much needed recreational focus for the ship's crew. The Chief Engineer, J D Morrison, had a flair for verse writing whilst Doorly was able to compose tunes. Together they wrote seven songs, using the strong voice of Teddy Evans to try them out on the crew. One of the most touching was the *Morning* 'hymn' whose chorus went:

> Far away in that cold, white land,
> In the home of the Great Ice King,
> Braving his fury, daring his wrath
> When honour and glory are showing the path;
> God will keep them from harm and scathe –
> Till the 'Morning' comes with the Spring.

INTRODUCTION

Doorly had already written the 'Voyages of the *Morning*' in 1915. After the war he started writing again, publishing short articles on various incidents from his voyages in *The Herald*, of Melbourne, and the *Bulletin*, of Sydney. He gathered these together and added others, publishing the collection as 'In the Wake' around 1936 in both London and Melbourne. A rare pamphlet, 'The Songs of the *Morning*', which contained all the tunes he had composed during the Antarctic voyages, was published by the Bread and Cheese Club in Melbourne in 1943.

Colbeck, the captain of the *Morning*, never published a detailed account of the voyages of the *Morning*, contenting himself with a three-page report in the *Geographical Journal* in 1904. Rather more information on the voyages is to be found in E R G R Evans's autobiography 'Adventurous Life' (Hutchinson *c*. 1946).

'Voyages of the *Morning*' is now a very scarce book. The story of the relief voyages is an important part of the history of Scott's expeditions. Here now is a new edition for those fascinated by the Heroic Age.

D W H WALTON
Bluntisham, 1995

THE VOYAGES OF THE 'MORNING'

Frontispiece

THE 'MORNING' IN McMURDO SOUND

THE VOYAGES
OF THE 'MORNING'

BY
CAPTAIN GERALD S. DOORLY, R.N.R.

WITH ILLUSTRATIONS AND A MAP

LONDON
SMITH, ELDER & CO., 15 WATERLOO PLACE
1916

[All rights reserved.]

TO MY CHUM

INTRODUCTION

WHEN Sir Clements Markham's life was cut short by the unhappy accident of Sunday, January 30, 1916, he was on the point of writing an introduction to the present book. The proofs had reached this office on the Saturday afternoon, and were to have been sent him on the Monday.

No one else could so appropriately have written a word of introduction. Sir Clements had a friendly hand for every Polar explorer, and his friendship, once given, was true and lasting. It was through him that 'The Voyages of the *Morning*' sees the light to-day, just as it was through him that the good ship herself was sent to relieve the *Discovery*. To him, as the friend and guide of the expedition and all concerned in it, the MS. was sent last summer, and on July 19, 1915, he wrote to Mr. Reginald Smith :—

'I have just received a narrative of the two voyages of the *Morning*; from New Zealand.

INTRODUCTION

The author was one of the junior officers—Gerald Doorly. He now commands a steamer out there, but he was a boy when he was in the *Morning*. The narrative is pleasantly written—first impressions of a keen observer. He was on board the *Discovery* and tells how Scott was constantly looking after the sick—bringing them tit-bits, &c. The whole story is well told. It occurred to me that it might form a sort of supplement to Turley's admirable book' [*i.e.* ' The Voyages of Captain Scott,' by Charles Turley].

In this sense, at least, he felt that the book would not be belated, ' just as a boy's book and an uncommonly good one—capital book for a prize,' though, owing to the pressure of a busy life, it appears long after the events it records. The boy is the eternal lover of adventure, and in this guise, moreover, lives on in many a grown-up. Assuredly there will be many to welcome such a record as this, for the story of the *Morning* has never been told, and it is worthy of being told to complete the tale of Antarctic adventure and the cycle of Scott's own travels.

L. HUXLEY.

March 1916.

PREFACE

MANY stirring accounts have been written during the last few years of voyages to the South Polar Regions—stories filled with fine geographical and scientific records, won by gallant deeds, unselfish devotion and, alas, also at the expense of valuable lives.

The voyages of the *Morning* as relief expedition to the *Discovery*, 1902–1904, have hitherto remained unpublished and the object of this book is therefore to fill the gap. The *Morning* was commanded by Captain William Colbeck, R.N.R.—an able and skilful seaman —to whom the greatest credit is due for the complete success of the *Discovery's* relief expedition. He was held in the highest esteem by the authorities, and by the late Captain Scott, and those who sailed with him have none but the happiest memories of his splendid leadership.

PREFACE

Commander Evans, C.B., R.N., who gained distinction in Captain Scott's last glorious but fatal expedition, was an officer on the *Morning*, and it was mainly through his affectionate endeavours that I became a member of the Relief Expedition. We started our sea career together on the training ship *Worcester* on the Thames, and this narrative is partly in the nature of a record of our many years friendship.

On the completion of the *Morning's* voyages, Evans was urged by Sir A. Conan Doyle to write and publish an account of the Relief Expedition, and he asked me to join him in this work. Circumstances arose, however, which prevented the arrangements being then carried out. I have always felt that the story of the Relief Expedition should be on record, and ten years having now elapsed since its conclusion, I have endeavoured to tell it before the vividness of the experience fades from the memory.

I last saw Evans in Lyttelton two years ago invalided from the Antarctic. He had spent the winter in the South, and in the sledging season following he was Captain Scott's

PREFACE

right-hand man in carrying out the work of laying depots and supporting the main party. He left Captain Scott in latitude 87° 35′ S., and was the last man to see the ill-fated polar party alive.

Every generation will send out its expeditions, journeying to the grim and lonely Antarctic. The adventurous men of future days will sail the icy seas we sailed upon, and gaze with fascinated interest at the great white lands, awe-inspiring and silent. Far away on the western shores of McMurdo Sound, they may see rising from the glacier-filled valleys, Mount Evans and Mount Doorly —a monument to a close friendship and affectionate association of those after whom they are named.

I tender my thanks to Mr. J. D. Morrison, chief engineer of the *Morning*, for supplying the photographs illustrating the story, to Captain D. Wilson-Barker, R.N.R., for the photograph of the *Worcester* and his interest in the welfare of his old cadet, and to the Royal Geographical Society for the accompanying map.

I especially extend my appreciation to

PREFACE

Mr. T. W. Whitson and Mr. Malcolm Ross, F.R.G.S., for friendly and valuable advice in regard to the publication of this book.

Last, but not least, I am deeply indebted to the late Sir Clements R. Markham, K.C.B., F.R.S., for his ever ready assistance and generosity. His memory is an inspiration to those who knew him. Through his kindly interest in, and encouragement to us all, he was truly named by the gallant Captain Scott :

'The Father of the Expedition and its most constant friend.'

G. S. D.

DUNEDIN,
 February 1916.

CONTENTS

CHAPTER I

H.M.S. *Worcester*—My early associations with Evans—Ambition to go to sea together—Boyish incidents—Obtaining the *Worcester's* two most coveted prizes 1

CHAPTER II

Temporary separation—Merchant Service and Royal Navy—A happy meeting—Astonishing telegram to join Antarctic Relief Expedition—First visit to the *Morning*—Difficulties in being appointed—Interviews with Captain Colbeck and Sir Clements Markham—Suspense—Meeting Mulock—Evans pleads our cause—Both appointed to Expedition . . 13

CHAPTER III

Delight in new work—Preparing for the voyage—Endeavours to obtain a piano—Doubtful hopes—Piano ultimately supplied by Sir Clements—Departure from East India Docks—Cheering ships—Sailing down Channel—Farewell to England 29

CONTENTS

CHAPTER IV

Bay of Biscay—Unique experience of a piano—Morrison, a lyric writer—Coaling at Madeira—Regrettable result of ' chain lightning '—The Equator—Court of Neptune . . . 35

CHAPTER V

First taste of bad weather—Run across the Southern Ocean—Concerts and nigger minstrel entertainment—Sighting New Zealand—A beautiful dawn—Arrival at Lyttelton—Hospitality of New Zealanders—Preparing for the Antarctic—Sailing day—Humorous incident at farewell service—An impressive send off 43

CHAPTER VI

Southward ho !—The stormy Southern Ocean—First experience of ice—The midnight sun—Beauties of Antarctica—Christmas Day—New island discovered—Ship strikes an outlying rock—Captain's skilful seamanship—The ice pack—Seals—Exercise and sport—Dangerous gale on edge of pack . . . 56

CHAPTER VII

First glimpse of polar continent—Cape Adare—Plans of Relief Expedition—The humorous penguins—' Southern Cross ' Expedition hut—First *Discovery* record found—Along the coast of South Victoria Land—Immense tabular bergs—Possession Islands—Coulman Island—Impenetrable ice pack—Futile efforts to reach Wood Bay—Compass error—Strange ski impressions—Desperate attempt to reach Franklin Island—Providential escape from insetting pack 68

CONTENTS

CHAPTER VIII

Difficulties attending our undertaking—Cape Crozier and Mount Terror—Depots to be established if no record found—Party at Cape Crozier—*Discovery* located—Delayed a week in heavy pack—Mount Erebus—The deadly stillness—Beaufort Island—A sailor's venture—Subsequent humorous sequel—Exciting steam up McMurdo Sound—Sighting *Discovery's* masts 80

CHAPTER IX

Ten miles of field ice—Experiments with ice-saws—First visitors—Return of sledge parties—A curious sight—Return of Captain Scott's southern party—First visit to the *Discovery*—The winter harbour—Meeting Scott, Wilson, and Shackleton—Banquet!—Captain Scott's speech—Astonishing hunger after sledging—An incident both humorous and pathetic . 96

CHAPTER X

Weather-bound—Sluggish ice field—Differences in seasons—Sledging over supplies—A coal depot—Sound freezing over—Farewell dinner on *Morning*—Modest requests—*Discovery* held for second winter—Departure of the *Morning*—A sad parting—'The Ice King'—Heavy new ice—*Morning* nipped—An anxious night—Our narrow escape—Sailing north—Days close in—Unpleasant weather—Arrival at Lyttelton 111

CHAPTER XI

Winter in Lyttelton—Expedition taken over by Admiralty — *Terra Nova* purchased — Incidents, humorous and otherwise — H.M.S.

CONTENTS

Phœbe assists in *Morning's* overhaul—Second departure from New Zealand—Tasmania—*Terra Nova* at Hobart—A unique tow—Transhipping stores—Sixty-third anniversary of *Erebus* and *Terror* at Hobart—Hospitality—An original dancer—Scuffle with a sailor—H.M.S. *Royal Arthur* supplies gun-cotton—Departure from Hobart 125

CHAPTER XII

Southward—*Terra Nova* and *Morning* in company—A narrow shave—The ice pack again—Second Antarctic Christmas—Scott Island sighted — 'Dead reckoning' — Admiralty Range—Deceptive distances—Fascination of the Antarctic—A quaint group—Sea-leopard hunt—View of South Victoria Land—Franklin Island—Difference in pack distribution—McMurdo Sound—Sighting *Discovery*—Extensive sheet of field ice—Hopeless prospects . 141

CHAPTER XIII

Captain Scott and Dr. Wilson—Astonishment at two ships—Doubtful chances of freeing *Discovery*—Evans and I at Scott's camp—An appetising 'hoosh'—Charm of an apple—Dangerous moving pack—Preparing to abandon *Discovery*—Sledging valuables to relief ships—Anxious days—Blasting operations—Snow blindness—Sailing orders issued—Depressing thoughts—Gale in Ross Sea—Extensive break-up of field—Visit to *Discovery*—Influence of swell at *Discovery*—Strenuous blasting—Sudden break-up of ice—Exciting rescues—Stirring hours—A struggle for priority—Dramatic arrival in winter harbour . 153

CONTENTS

CHAPTER XIV

Preparations for freeing *Discovery*—*Morning* adrift — Terrific explosions — *Discovery* released—Thrilling scene—A severe blizzard—*Discovery* driven ashore—Heavy weather—Trying experiences—Ships alongside glacier snout—Transhipping coal and stores—' Follow me '—Start of homeward voyage . . 173

CHAPTER XV

A beautiful Antarctic scene—Ships in company—Impressive coast-line—*Morning* parts company — *Terra Nova* escorts *Discovery* — Boisterous Southern Ocean—Engines break down—An anxious month—Reverses making Auckland Island—Arrival Port Ross—A peaceful contrast—*Discovery* and *Terra Nova* at rendezvous—Pleasant days—Ballasting ship—News of outer world—Departure from Auckland Island—A fine slant—Return of National Antarctic Expedition . . . 188

CHAPTER XVI

A great welcome—Evans's wedding—Departure for England—Around Cape Horn—The Falkland Islands—Last run of voyage—Depressing head winds—Change off Cape Finisterre—Running before a gale—The dear Homeland—*Morning*, steamer or sailer ?—Plymouth Sound—The end of the cruise . . . 201

CHAPTER XVII

Morning laid up at Devonport—Our diminutive floating home—Dividing the ' spoils '—Crew paid off—Sailors' affection for their ship—The last of the *Morning*—Homeward-bound crew—Our unconscious humorist—The Great City—Home, sweet home 210

CONTENTS

CHAPTER XVIII

Expedition functions—Captain Scott's lecture at Royal Albert Hall—Graceful tribute to Captain Colbeck—Reception of *Morning's* crew at Hull—Royal Corinthian Yacht Club's dinner—Captain Colbeck's wedding—The sad farewell 217

ILLUSTRATIONS

THE 'MORNING' IN MCMURDO SOUND	*Frontispiece*	
H.M.S. 'WORCESTER'	*Facing page*	8
MORRISON'S SCHEME—'AT ONCE SURPRISING AND ORIGINAL'	,,	36
THE AUTHOR	,,	42
EVANS AND ENGLAND	,,	44
OBSERVATION HILL	,,	59
PART OF SCOTT ISLAND AND HAGGITT'S PILLAR	,,	59
AN ANTARCTIC PILLAR-BOX	,,	74
DELAYED FOR A WEEK	,,	74
CAPTAIN COLBECK READING 'DISCOVERY' RECORDS FOUND AT CAPE CROZIER	,,	84
'EVERY MOVEMENT WAS KEENLY WATCHED'	,,	98
APPROACH OF OUR FIRST VISITORS	,,	98
CAPTAIN SCOTT LEAVING THE 'MORNING'	,,	104
'DISCOVERY' IN WINTER QUARTERS	,,	104
SLEDGING GEAR AND SUPPLIES TO 'DISCOVERY'	,,	112
COAL DEPOT ON GLACIER TONGUE	,,	112

ILLUSTRATIONS

SHACKLETON ON ARRIVAL AT LYTTELTON.	*Facing page*	126
'ENGLAND'S DOWNFALL'	,,	134
CAPTAIN SCOTT AND CAPTAIN COLBECK	,,	154
CAPTAIN SCOTT'S TENT, CAPE ROYDS	,,	158
ARRIVAL OF MAIL FROM 'DISCOVERY'	,,	158
AN OCCASIONAL INCIDENT—ADRIFT ON THE SEA ICE	,,	168

MUSIC

SOUTHWARD	*Between pages*	38, 39
THE ICE KING	,,	116, 117

MAP

SHOWING TRACK OF THE S.Y. 'MORNING'		*At end of text*

THE VOYAGES OF THE 'MORNING'

CHAPTER I

H.M.S. *Worcester*—My early associations with Evans—Ambition to go to sea together—Boyish incidents—Obtaining the *Worcester's* two most coveted prizes.

WHEN passing up or down the Thames between Erith and Gravesend, one cannot help noticing a number of old wooden fighting ships moored on each side of the river. A century ago these were the pride of the British Navy, and upheld the glory of the nation in many conflicts and on many seas. Now they are spending a peaceful old age as training-vessels for the British youth of every class.

Off Greenhithe lies the *Arethusa*, used as a training-ship for poor boys of good conduct.

THE VOYAGES OF THE 'MORNING'

These are generally drafted into the Royal Navy and become seamen.

About a mile distant, and moored off the historic Ingress Abbey, lies that excellent training-ship 'for young gentlemen desirous of becoming officers in the Mercantile Marine,' known as the Incorporated Thames Nautical Training College, H.M.S. *Worcester*.

The *Worcester* and her complement, the *Conway*, on the Mersey, during their many years of existence as training-vessels, have turned out some of the finest men; and in almost every corner of the globe 'old *Worcesters*' and 'old *Conways*' are to be found as captains of great liners, officers of ships of all descriptions, while not a few hold prominent shore appointments in various parts of the world. A certain percentage have also passed into the Royal Navy through the Admiralty granting several nominations from these ships annually.

Boys from all parts of the world have been trained on the *Worcester*, hailing not only from the British Possessions, but from many foreign countries as well, so that it is not to be won-

ON THE 'WORCESTER'

dered at that old boys are to be found in the most unexpected places.

A number of years ago now, two boys—both under fifteen years of age—joined the *Worcester* as cadets within a few months of each other. One, the son of a barrister practising in London, had joined the ship for the purpose of passing into the Royal Navy. The other was the son of a clergyman, and hailed from Trinidad, in the British West Indies. The former was Evans, affectionately known as ' Teddie,' and the latter was the writer of these notes. Coming as we did from opposite corners of the world, so to speak, we met as strangers knowing nothing of each other; but it happened that we were placed in the same ' top ' (or section), and slept in hammocks slung next to one another. Boys thrown together in such a way soon get to know each other favourably or otherwise. In the course of a very few weeks we became inseparable chums, and, as boys do, swore eternal friendship. My memory takes me back to the many happy times we spent in each other's company. During holidays we frequently exchanged visits, and our boyish escapades were

THE VOYAGES OF THE 'MORNING'

quite in keeping with a certain spirit of adventure which is so dear to most boys, and particularly so to the budding sailor, who holds ever before him the ideal of endeavouring to live up to the great and glorious traditions of the sea.

Evans was a sturdily-built boy, heavier, but a trifle shorter than myself, and was a splendid muscular specimen, whilst I belonged to that type known as wiry. We used to put in many a half-hour at horse-play, which, speaking for myself, was very hard work. Our wrestling bouts were terrific struggles, and although I was frequently the loser, I think that Evans will admit that for a light-weight I put up a very good fight, and made him exercise his muscles to their utmost. We ran about equally well, and have gone for miles together, each, no doubt, not caring to give in to the other. We were also well up in gymnastic and athletic feats.

I remember on a holiday once riding a tandem bicycle with him through some of the busiest thoroughfares in London, oblivious of the fact that he had never been on a bicycle before. He stated that he would supply all the ' horse-power ' necessary if I did the more scientific work of steering and ringing the bell. We rode for miles, and, marvellous to relate, escaped accident.

In boyish fashion we often competed in odd feats of endurance. One wet evening, after a performance at the Gaiety Theatre, there was the usual rush and scramble for 'busses. Evans's parents were with us, and they succeeded, after several reverses, in getting standing room only. As the stream of packed 'busses rolling by offered little prospect for us, we decided to run home. I am not sure how far it is from the Strand to Regent's Park, but in the bleak, muddy night it seemed quite a long way. We stuck to the game, overtook the 'bus, and had a cheerful fire going and supper ready to greet the others.

Evans stayed with me one night. We arrived late, and found that the family had gone to bed and had locked us out. We did what any active boys would do, I suppose, in like circumstances. By standing on his shoulders I wriggled through a dense creeper, shinned up a drain-pipe, and reached the

THE VOYAGES OF THE 'MORNING'

bedroom window. My brother awakened with a start, but, being used to our playful ways, he turned over with a grunt and was soon snoring again. I let Evans in respectably by the front door, but, in spite of stealth, the stairs creaked horribly.

We were at Margate one Easter, and went out for a pull on a bitterly cold day. To my surprise, when we were some distance from the shore, Evans suggested a swim. I shuddered at the thought, but I couldn't be outdone by my chum. We stripped and dived in. It was awful; Evans was enthusiastic over its 'loveliness'; I agreed with a gasp, but was never more thankful than when we were dressed and, with chattering teeth, were pulling back to the pier.

We made a boat excursion one day from Ramsgate to the Sandwich River, a few miles away. Evans' mother and brothers accompanied us. Evans and I considered ourselves expert seamen, but we had a lot to learn. We ran the boat into a creek off the main stream, and, as it was raining, we improvised a tent with oars and sail.

STRANDED ON THE EBB

There was nothing to do but to dig into the provisions. We became so engrossed in this pleasure that the tide gradually ebbed away and left us stranded in a slough of mud: 'Women and children first,' of course; in a few moments feet were bared, and, sinking to the knees in mud, we gallantly landed the lady of the party. After some bargaining at an adjacent inn, we obtained a trap, in which she was ignominiously despatched in the approaching night to Ramsgate. It was impossible to move the boat from the muddy bed, so we decided to 'stand by' and keep 'watch and watch' through the gloomy night—for what reason I cannot remember now. We had a hazy idea, I think, that when a vessel was stranded it was necessary for the crew to suffer certain hardships and discomforts. Fortunately for us some coast-guardsmen happened to pass by early in the evening. When we explained our predicament they laughed good-humouredly and told us to go home. They offered to look after the boat when the tide made enough to refloat her. As our enthusiasm had been on the wane, we gladly accepted their assist-

THE VOYAGES OF THE 'MORNING'

ance, and being cold and wet, we did a five-mile non-stop run home.

We found the boat afloat next morning, and with pangs of anxiety concerning the owner's feelings, we sneaked back with humiliation to Ramsgate. An unexpected welcome awaited us. Another *Worcester* boy—possessed of an extraordinarily inventive mind—was on the pier surrounded by a crowd of trippers. Quite unusual interest seemed to be centred in our arrival at the steps. We soon heard that this imaginative wag had spread a yarn that a battleship was wrecked in the Downs, and that we were alleged to be two midshipmen who had effected a daring escape! Our *Worcester* uniforms lent the necessary colour to this absurd story.

Among the many quaint customs on the *Worcester*, is a recreation known as 'slewing,' which comprises walking round and round the upper or main decks, arm in arm with any particular friend during 'stand easy' times. It was amusing to see, perhaps, 150 boys walking round and round the decks in this fashion in twos chiefly, but occasionally in threes. There was always a proper

H.M.S. 'WORCESTER'

'SLEWING'

decorum to be adhered to before 'slewing,' very similar to requesting the pleasure of a dance in a ballroom. It was not conducted quite as politely, perhaps, though the 'Come and slew' produced a more or less similar effect.

Needless to say, Evans and I put in many an hour's 'slewing' during our terms on the ship, and, as may be imagined, all our hopes and ideals were discussed, and it can safely be stated that no two boys ever knew or understood each other better than we did. At one period, our ambition was to go to sea together, and we speculated as to its likelihood. One arrangement was that if he failed to pass into the Navy, we should apprentice ourselves to some sailing-ship firm; but much as I should have appreciated this, I always felt that his heart was set upon the Navy, and consequently encouraged him in that direction.

I recall to mind most vividly one evening towards the end of our stay on the *Worcester*, just before prayers and 'pipe down' (turn into hammocks). We were 'slewing' together as usual. Evans, who was working hard, and doing many extra hours for the Naval Examination, seemed rather despondent. He was continually harping on the subject of the examination and its consequent results—dreading failure. There were seven or eight other *Worcester* boys to compete with him, and a similar number from the *Conway*, I believe, out of whom perhaps only the first two or first three would get through. I felt in my own mind that Evans was good enough for any of them. The hard work and long hours had doubtless caused him fatigue and anxiety. I cheered him up as much as I was able, and assured him that I had confidence in him and felt certain that he would pull through swimmingly.

Now, at the after end of the main deck on the *Worcester* are two large panels under glass, one on either side. One contains the Ten Commandments printed in gold lettering on a black background, and the other has, also printed in gold, the list of the cadets who have obtained the *Worcester's* two most coveted prizes since the establishment of the ship as a training vessel. These are the two prizes granted by Her late Majesty

LINKED NAMES

Queen Victoria, and continued by the late King Edward, and by King George. The first is the Queen's Gold Medal, awarded annually to the boy who shows qualifications likely to become the finest sailor, decided by a ballot of the cadets; and the second is the cadetship into the Royal Navy. The two prizes are regarded as ideals, and I don't suppose there has been a boy who has passed through the *Worcester* who has not known, almost by heart, the printed regulations on this board, and the names of the successful prize-winners, printed in gold, as far back as 1869.

At each turn of the deck we made on this evening, these printed records confronted us, and Evans, after gazing at the prize board for some moments, suddenly remarked : ' I wonder if my name will ever be printed there.' I replied that I considered it quite likely, whereupon he squeezed my arm and said, rather impulsively, ' Dear old chap, I'll wager that if mine is put there, yours will be printed on the opposite side.' This compliment came as a wonderful surprise to me, as I had never in my wildest hopes ever

THE VOYAGES OF THE 'MORNING'

thought that I should have the remotest chance of being selected even to be voted for, or against, by my shipmates. Evans had no idea either; how could he? But he felt that it would be a fitting conclusion to our career on the dear old *Worcester*.

A month or two passed, and the results of the Naval Examination being published, it was seen that Evans had passed very creditably first, above all competitors from both ships. A few more months passed, and I was voted 'Queen's Gold Medallist.' Evans was indeed prophetic: our names are printed in gold abreast each other on that coveted scroll of fame. It was an impressive end to our training-ship days, and in keeping with our close friendship.

CHAPTER II

Temporary separation—Merchant Service and Royal Navy—A happy meeting—Astonishing telegram to join Antarctic Relief Expedition—First visit to the Morning—Difficulties in being appointed—Interviews with Captain Colbeck and Sir Clements Markham — Suspense — Meeting Mulock — Evans pleads our cause—Both appointed to Expedition.

As the steam vessel has largely superseded the now almost obsolete sailing ship, few are the openings left for sailors to pick up a satisfactory freight. The chief line of business left to sailing ships appears to be carrying coal from West of England ports to the West Coast of South America, and, by way of securing return cargo, loading nitrate of soda on the Chilian coast, or proceeding in ballast to the Western States of America for wheat.

During my apprenticeship ports on the West Coast of the two Americas were visited, and it was at Antofogasta in Chili, while I

THE VOYAGES OF THE 'MORNING'

was engaged in the dirty but unavoidable occupation of shovelling coal, that a letter was thrown to me in the hold. It bore a Maltese stamp and postmark, and the writing was unmistakably that of Evans. We were fast friends, yet what a contrast was there in our callings! We were both becoming seamen, but by very different methods.

At this time Evans was a midshipman on a first-class cruiser in the Mediterranean Squadron. His letter was not written in a cheerful strain; he complained of being unhappy and discontented, and that 'for two pins I'd chuck the Navy, and apprentice myself with you.' Now, I could not claim much worldly knowledge, and comparatively little of the sea, but it was quite clear to me that whatever Evans's life might be, it surely was not as unpleasant as mine. Reflect upon it as I would, I could not understand how he could willingly desire to exchange a position in the service of his country for the back-aching, rough-and-tumble existence that I was carrying on.

As perhaps he did not realise the nature of my life as well as I could picture his, I

A SUFFERING MIDSHIPMAN

wrote off immediately urging him to stay where he was, and not for a minute to entertain such a foolish idea. I learned later on that Evans was just passing through that juvenile stage in the Navy where one has to be subservient to the officer told off to keep the middies up to the mark, and the officer in question was evidently tactless and generally unpopular with the midshipmen. Evans, being physically the officer's superior, strongly resented his alleged bullying, and would willingly have thrashed him were it not that such insubordination would most probably result in dismissal from the service.

Whether my letter played any part in influencing the irate midshipman or not, I never really enquired; suffice it to say that he did not resign from the Navy, and the next time I heard of him he had become the life and soul of his ship, and was considered one of the smartest and most popular midshipmen on the station.

After the fashion of boys, we kept in touch with each other by corresponding about once a year; but we frequently heard of each other. Whilst in San Francisco on one occasion, I

THE VOYAGES OF THE 'MORNING'

heard with regret that poor Evans had been dangerously ill with fever, and that, after being laid up at Malta until convalescent, he was invalided home. Except that I knew of his subsequent recovery, I lost track of him for a brief period.

Several years had now passed, and at the termination of one voyage I remained on shore in order to undergo some training in the Royal Naval Reserve on the drill ship H.M.S. *President*, lying in the West India Docks, London. I had received a nomination as midshipman in the R.N.R. on leaving the *Worcester*.

Residing at that time near Blackheath, I travelled to and fro daily to drill. On returning from drill one afternoon, just as my train was drawing up at London Bridge station, I noticed another train moving out. My surprise and delight knew no bounds when I suddenly caught sight of my old pal waving frantically out of the carriage window of the departing train.

There was no means of communication, but he pointed wildly to the superscription 'Greenwich' on his train, indicative of his

REUNION

whereabouts. I formulated a plan for the next day, and after drill returned home via Greenwich, and called at the R.N. College.

Our reunion was emphasised by a warm embrace. There was so much to tell that it was difficult to know where or how to begin. We went to his room, which he proudly informed me was a post captain's. He had been rather run down in health, and having assured the Captain of the College that the Sub-Lieutenant's quarters were damp and draughty, he was accordingly allotted this luxurious apartment.

We arranged to meet ' under the clock ' at Charing Cross that evening, and from 7.30 P.M. until 10.30. P.M. we dined, or, more correctly, talked, with intervals for food and drink, at a restaurant in the vicinity of Leicester Square, of which Evans was an occasional patron. A lady and gentleman were sitting on the opposite side of our table when we arrived, sipping their coffee and liqueurs, and it was evident that our breezy and animated conversation must have completely absorbed their attention, because they didn't leave until we did, and had consumed, for appearance's sake presum-

THE VOYAGES OF THE 'MORNING'

ably, innumerable cups of coffee and glasses of liqueurs.

We met occasionally after this. Evans was working for his lieutenant's examination, and a great deal of his time was occupied in study. Before I went off to sea again he had passed the first two sections of his examination, obtaining a 'one' in each.

I was now a junior officer in the P. & O. S. N. Co., and had been appointed to a transport vessel, temporarily fitted and employed as a hospital ship during the South African War. The voyages were from Southampton to the Cape, conveying detachments of troops to the Front, and returning with a pathetic human freight of sick and wounded soldiers. Arriving at Southampton one voyage, about the middle of June 1902, a great surprise awaited me.

Amongst my letters was a telegram from my friend Evans, containing the astonishing query: 'Would you care join Antarctic Relief Expedition? Ship *Morning* sailing early July. Friends quite agreeable. Reply at once or place will be filled.'

I went to my cabin to consider this most

FIRST SIGHT OF THE 'MORNING'

unexpected and almost startling proposal. Why a relief expedition ? To relieve whom ? It dawned upon me that it must necessarily be connected with the *Discovery* Expedition, which had sailed twelve months before for the Antarctic. But what did I know of such things ? Was I suitable for such an undertaking ? Then it suddenly flashed across my mind that by joining this Expedition the greatest wish of our lives would be gratified. Evans and I would at last sail in the same ship, unusual and extraordinary though that ship and voyage might be.

My mind was soon made up, and, obtaining leave from the captain, I proceeded to London. My thoughts and feelings on the subject were naturally somewhat mixed, as I was working more or less in the dark; the only satisfaction I seemed to feel was that, as Evans was there, all must be well.

Proceeding directly to the East India Docks, I found after a little search and enquiry the object of my quest, and I have to confess to a deep sense of disappointment.

The ship I had just left at Southampton was a P. & O. liner of some pretensions;

the eye was accustomed to such dimensions; but the object I now gazed upon seemed so ridiculously insignificant that for a moment I wondered if there hadn't been some mistake. It was only too true, however, that this small barque-rigged vessel of 290 tons, with a quaint little yellow funnel in the after-end, and a strange-looking tub at the mainmasthead, was the steam yacht *Morning*, in the course of preparation for her adventurous voyages.

As there was no visible sign of life about the ship, it appeared to me that my hurried journey to the docks had been unnecessary, and I was deliberating as to my next move, when up popped my friend's head from the one absurd little hatch the *Morning* possessed, and seeing me he gave vent to an outburst of joy with his 'Hello, Jose' (a fancy term of endearment). 'Welcome to the Dreadnought.'

I wondered at the zeal and enterprise of my old chum. About three months previously he was going through the torpedo and gunnery course at Whale Island, when one day, during 'stand easy,' he chanced

A THIRD OFFICER WANTED

to read in a daily paper that the Arctic whaler *Morning* had been purchased from a Norwegian firm in Tonsberg, and after overhauling and equipping her in London, it had been decided that she would sail for the South Polar regions as a relief ship to the *Discovery*. The energetic Evans proceeded at once to London and personally interviewed Sir Clements Markham, then the President of the R.G.S., with a view to joining the Expedition. He was introduced to the captain, who recommended him there and then as second executive officer. His good physical appearance, his frank and zealous manner, and his enterprising offer to man the ship with blue-jackets from Portsmouth if necessary, no doubt appealed very strongly to the organisers.

As the time drew on, and it was necessary to consider the appointment of a third executive officer, Evans, being by this time much thought of by the authorities, suggested emphatically that they could not do better than appoint me to the position!

This, as will have been seen, was done totally without my knowledge; but knowing me as he did, and remembering our early desire to go to sea together, he was positive that I would not only accept the position, but would be a keen and useful officer to the Expedition. This was indeed the acme of friendship. There were already over fifty applicants for the position, but by the time I arrived on the scene, through Evans's enterprise and affection, these had been reduced to a possible two, of whom I was one. The other applicant was a young naval officer, Mulock, who had surveying experience amongst his recommendations, and was altogether on a better footing than I was for the appointment.

Evans was perplexed as to how the decision would go, for although extremely keen for me to join, he was nevertheless generously disposed toward his brother officer, who was in all respects equally suitable, and by his credentials appeared to be even better qualified than I was.

I met Captain Colbeck on board. Though quite a young man, he impressed me as being the ideal type of commander; and, as I subsequently learnt, he had already

SIR CLEMENTS MARKHAM

experienced an Antarctic voyage, being one of the first party to spend a winter in the frozen south. He explained that Evans's telegram to me was rather premature, there being much difficulty in deciding on the officer to be appointed ; but in spite of this, he held out a hope for me, and even went so far as to describe the nature of my duties, in the event of my being appointed. This sounded cheerful and encouraging, but there was nevertheless the uncertainty, which gave rise to conflicting emotions. He promised to see Sir Clements on my behalf that evening.

Returning to town in this unsettled frame of mind, I decided to anticipate the captain, and called personally on Sir Clements. As a boy on the *Worcester* I remembered Sir Clements coming on board occasionally during the winter terms and delivering very interesting lectures to the cadets on Arctic exploration. One of our old boys was at that time in the Arctic, with the Jackson-Harmsworth Expedition, and Sir Clements himself had been, as a young naval officer, on one of the search expeditions

THE VOYAGES OF THE 'MORNING'

to the Arctic in quest of the fatal Franklin Expedition.

Somehow I felt that in Sir Clements I should find a friend, and I knew that he took an interest in the welfare of *Worcester* boys. The interview could not have been more pleasant, although no decision was arrived at regarding my appointment.

I was astonished to learn how much he knew of me and my record, and it was quite clear that my good chum had painted me in the most favourable colours. The difficulty of making a selection from so many apparently excellent men was pointed out, but I left convinced that my only rival was the young naval officer before mentioned. I was also informed that the decision would be conveyed to me in a few days' time.

Now, although Sir Clements had communicated with the P. & O. Company about granting me the necessary leave should I be appointed to the Relief Expedition, I found on calling at the Company's office that I had been promoted, and appointed to a ship sailing in a few days for Bombay. On enquiring as to their knowledge of my

SUSPENSE

joining the Expedition, it was stated that such communication had been received some time before, but as no information on the matter was forthcoming from me, it was taken for granted that I had either been unsuccessful or had decided not to go, and had been promoted in the ordinary course of events.

As a matter of fact, as I had only that very day known of the possibility of going in an expedition, I could hardly have communicated earlier. Seeing that my chances were favourable, I applied for, and was granted, the necessary leave. This was perhaps what one might term a sporting stroke on my part, but had I sailed away to Bombay my even doubtful chance would have been lost.

One long anxious week passed by, and no information was forthcoming. The suspense was trying, for the fascination of the adventure was gripping me, and I was in that frame of mind that to join the Expedition was my greatest ambition, even were I to serve in any subordinate position.

Saturday arrived, and I could contain

myself no longer. To wait over the week-end seemed an eternity. Hiring a bicycle, I rode with all haste through Blackheath and Greenwich, and finding a short route through the Blackwall tunnel, soon arrived at the *Morning*. A scene of much activity now centred in the little ship; running rigging was being rove off, sails bent, accommodation fittings placed, extra houses and lockers being constructed, and dray-loads of stores and gear were continually arriving and being stowed away expeditiously.

Evans was in the thick of it, and I longed to be giving him a helping hand. I had hoped that he would be able to tell me the best, or worst, regarding myself, but he knew nothing. I explained everything that I had done in the matter, and there did not seem to be anything more to do.

Whilst chatting on the little poop-deck, Evans, who was looking towards the dock gates, suddenly exclaimed, 'Well, here's the other man, Jose,' and in a few minutes my rival stepped on board. Evans introduced us, and although we assumed a nonchalant attitude, each had heard of the other. Mulock stated

TAKE BOTH!

that although he had not heard definitely, he was given to understand that he would be appointed, and that his commission was on board.

My chagrin may be imagined. Here was I within an ace of being accepted to sail with my best friend, and yet it certainly appeared as though all my fond hopes and air castles were to be ruthlessly demolished at the eleventh hour.

The position was discussed thoroughly, and each displayed such a keenness to go in the ship that Evans considered it pertinent to offer the suggestion that both of us be appointed. This was characteristic of him—somewhat impulsive and wholly original—but I did not for a moment entertain in my mind any likelihood of this scheme taking effect, for the reason that the Expedition funds were low, and every extra man must, of necessity, be an additional expense. Evans, however, was optimistic, and having invited us to his half-built cabin, he presented us with some liquid in a tin pannikin, out of which we drank in turns success to the Antarctic Relief Expedition.

THE VOYAGES OF THE 'MORNING'

This rite being duly performed, he stated that he himself would plead our cause to Sir Clements that very day, and as an afterthought, before leaving, in the event of both being appointed, he tossed a coin for the respective ranks of third and fourth executive, which fell favourably to me.

More or less jubilant, I returned home, but looking at the bare facts fairly, my chance of success did not seem hopeful. Just before turning in that night a telegram arrived for me. It read: 'Cheer up; both going. —TEDDIE.'

CHAPTER III

Delight in new work—Preparing for the voyage—Endeavours to obtain a piano—Doubtful hopes—Piano ultimately supplied by Sir Clements—Departure from East India Docks—Cheering ships—Sailing down Channel—Farewell to England.

IT is difficult to describe exactly one's feelings when some deep sorrow or great joy suddenly overtakes one, and the joyous impression made by that telegram is hard for me adequately to express. For a considerable time I had been living in a perplexing state of doubt—the last few days, indeed, had been almost unbearable—and now, literally at the eleventh hour, came this very gratifying information.

How it came about I did not at the time stop to consider; the main fact was that all anxieties and doubts were dispelled, and I was now to join the Expedition. Through it all, however, was the feeling of a deeper esteem for my good chum, without whom my appointment could hardly have been secured.

My duties commenced at once, and the work was absorbingly interesting. Evans and I were boys again, extremely happy, and the more we laughed and jested, the harder we worked. It was an ideal experience. We lunched together each day at a neighbouring hotel, and it was during these pleasant retreats that I learned the nature and extent of our undertaking. Evans was thoroughly conversant with the minutest details of the programme mapped out for our voyages, and in a very short time I was well posted in the intricacies of the proposed work of the Relief Expedition.

The last week was naturally a very busy one, the days being fully occupied, and the evenings devoted to many farewell functions which were especially entertaining and interesting. The day before sailing the equipping was practically completed, the only noticeable omission being a piano for the ward-room. There are some who would perhaps regard this item as an unnecessary luxury, but I can assure them that it is quite an important ad-

THE GIFT OF A PIANO

junct, in that it undoubtedly tends to improve the feeling of *camaraderie*, so essential in an undertaking of this nature, and it is acknowledged that most sailors are true lovers of music. As a piano was not included in the specifications, Evans and I conferred as to a means of obtaining one. Some ladies visiting the ship one afternoon remarked on this deficiency, and the situation being explained (which, by the way, we never forgot to mention to any likely donor), one girl, carried away with enthusiasm, said that there was a spare piano at her home which she thought we could have. This generosity was quite touching, but we subsequently heard that when the suggestion was proffered, this well-meaning lady's parents politely but firmly refused to entertain any such extraordinary idea! As the time drew on it seemed that our hopes in this direction were doomed to disappointment, until, the evening before departure, Evans volunteered, as a last resource, to approach Sir Clements on the matter. The night was far advanced when he arrived at Eccleston Square—so far, indeed, that Sir Clements was in the act of going to bed. A call at this late

THE VOYAGES OF THE 'MORNING'

hour naturally caused surprise, and Evans, with profuse apology, explained the object of his nocturnal visit. As a friend in need Sir Clements was pre-eminent, and, realising how keen we really were, he arranged to present us with his own spare piano. This kind and generous action was in keeping with the wholehearted interest he always displayed in the welfare and comfort of each member of the Expedition.

Early next morning a van was dispatched to convey the piano to the docks, but the day wore on and no piano arrived. Owing to a mistake in the address, or the stupidity of the driver, no van arrived at Sir Clements', and there was now no time for it to reach the docks before the ship sailed. Here, indeed, was a sad disappointment for us, and with small comfort and less effect, we hurled grave epithets at that carter's ancestry far back into the dark ages.

A crowd had now gathered on the dock side, a bravery of bunting streamed from a hundred masts, the last fond farewells had been taken, cheers were ringing, and whistles were tooting their messages of 'Good luck' and

THE START

'*Bon voyage*,' ' Stand by ' had rung, and the picked crew were assembling at their allotted stations, when a commotion in the crowd, which gave way, permitted a van to range hurriedly alongside with the jangling of harness and panting of horses. In a trice the back was thrown open, a piano revealed, and with many willing hands, this last item in the *Morning's* equipment was unceremoniously dragged on board. This was Sir Clements' farewell kindness. On learning that there was no hope of conveying his piano in time, he had despatched a servant with all haste towards the docks to purchase one at the nearest piano establishment, and this prompt and kindly action enabled the *Morning* to sail complete in every detail.

The day was not attractive as far as the weather was concerned, but the good cheer extended from every quarter was impressive. Steamer after steamer passed by, tooting their whistles and dipping their ensigns. Each training-ship sent up a ringing cheer as we passed, and the lusty cheering that greeted us as we approached the old *Worcester* was not likely to be forgotten. It was a proud moment for Evans and myself.

THE VOYAGES OF THE 'MORNING'

It is true that the sailing barges overhauled and occasionally passed us during our progress down the river, but what of that? They could not face the tempests of oceans, nor the rigours of Antarctica!

It is true also that the tide turning caused us to come to anchor off Gravesend, and next day, owing to a headwind, it was found advantageous to anchor again off Deal. But our enthusiasm enabled us to overlook such drawbacks.

A fresh fair breeze making, we eventually got under way, and setting the brand new suit of sails for the first time, the little ship bowled merrily down Channel, while we cheered and exchanged signals of good luck with ships of all nations. It was good to be alive!

Rounding to off Start Point, our pilot and the few good friends who had accompanied us thus far bade us a hearty farewell, the last cheers were given, and the *Morning*, heading for the open sea, sailed away into the unknown future.

CHAPTER IV

Bay of Biscay—Unique experience of a piano—Morrison, a lyric writer—Coaling at Madeira—Regrettable result of 'chain lightning'—The Equator—Court of Neptune.

THE Bay of Biscay enjoys a bad reputation, storms are of frequent occurrence, but for us it had no terrors, having donned its Sunday apparel of calm, sunny, and peaceful weather.

Owing to the more important work of getting everything ship-shape during these early days, the piano had been left in undignified association with various marine stores, ice-picks, and sledging equipment. Its turn of recognition, however, duly arrived, and an attempt was made to transfer this valuable asset to its permanent and more suitable environment. I use the word 'attempt,' because it was soon discovered, to our utter perplexity, that neither door leading down to the ward-room was wide enough to permit of its passage. Here was an obstacle which demanded much ingenuity to overcome. It was inconceivable to think that after all the carefully wrought schemes to procure this useful and ornamental article for the ward-room, it would possibly have to spend its days in the obscurity of the sail locker on deck.

The absurdity of our dilemma caused us, for a moment, to laugh outright. We were 'up against' a proposition, however, which even the resourcefulness of the 'handy man' failed to solve.

Leaving the piano on deck, we went below to lunch, during which every practicable (and impracticable) suggestion was put forward to overcome the difficulty, but to no avail. It was not until after the meal that Morrison, the chief engineer, who had remained unusually reticent over the matter, submitted a scheme which was at once surprising and original. Rather apologetically, he assured us that he was no musician, but after carefully examining the inner construction and mechanism of the piano, he could not see that the instrument would suffer any material detriment by being cut in half!

MORRISON'S SCHEME—'AT ONCE SURPRISING AND ORIGINAL.'

THE PIANO DISSECTED

This astonishing suggestion was received with wonder and incredulity, but as there seemed to be no other feasible solution, and as we had already begun to regard our chief engineer not only as a shrewd, but as an extremely practical and capable member, we eventually approved of his idea, and with much interest, witnessed his unique demonstration.

The keys were easily unshipped, and each side of the key-board was carefully sawn through—the cook's meat-saw proving the most useful implement by reason of its serviceable shape and general greasiness. The key-board was now detached from the main body of the piano, the parts passed down to the ward-room, and with the aid of some glue and a few wooden dowls, the key-board was once again neatly secured, the keys re-shipped, and, to celebrate the great achievement, a concert with the full crew was held that evening. That piano was a blessing not to be overestimated, and we often wondered how we should ever have done without it. On Saturday nights in fine weather concerts were held for the crew, hymns were sung on Sunday evenings, and many a sweet memory of home

THE VOYAGES OF THE 'MORNING'

and loved ones was awakened by a hundred familiar melodies.

The chief engineer, among many other accomplishments, proved to be a writer of more or less topical verses. Being of a musical turn, I arranged music for these, and my modest efforts at composition were, at any rate, a source of much pleasure to me. When one of these songs was to be inflicted upon the company, Evans always volunteered to give it what he termed 'a steam trial' by singing it, and, if the trial were successful, the song was allowed the freedom of the ship.

The following is the first song written, and may be of interest :—

SOUTHWARD.

As I fly on the wings of the *Morning* o'er the sunny tropic seas,
When laughing waves dance round the ship and there's love in the song of the breeze,
At night on deck, 'neath the starlit sky, when only the gods can hear,
I think of a girl in the far Northland, and I wish that my love were near,
And wish that my love were near me now, when only the gods can see,
Lazily loving the hours away on the sunlit summer sea.

Southward

Words by J.D. Morrison
Music by G.S. Doorly

1. As I fly on the wings of the "Morning", O'er the sun'ny tro-pic seas, when laughing waves dance round the ship and there's love in the song of the breeze, at night on deck 'neath the
2. As I fly with the wings of the "Morning", O'er the cold, dark ice-bound seas, when shouting waves rush past the ship and the stormfiend rides on the breeze, on the wave-washed deck 'neath the

HOURS AWAY ON THE SUNLIT SUMMER SEA
SOUTHERN SEA THROUGH THE BLINDING SLEET AND SNOW

1ST VERSE ONLY   2ND VERSE ONLY.

A SHIP'S LYRIC

As I fly with the wings of the *Morning* o'er the cold ice-bound seas,
When shouting waves rush past the ship, and the storm-fiend rides on the breeze,
On the wave-washed deck 'neath the murky sky, when only the gods can hear,
I think of a girl in the far Northland, and I wish that my love were near,
And wish that my love were near me now, when the westerly breezes blow,
Driving along o'er the southern sea through the blinding sleet and snow.

We were fast settling down to our new life, and becoming more or less accustomed to the limited capacity of our little vessel by the time we arrived at Madeira. This visit was decided on through a generous offer of a supply of coal by the Union Castle Company.

Our stay of three days was enjoyable, and Evans and I spent some happy hours on shore rambling about among the beautiful hills, rich in sub-tropical foliage. Having discovered the most pretentious hotel, we indulged in the luxury of a hot bath—on the ship only a comfortless necessity—after which we dined well and, I take it, wisely.

On the afternoon of our departure some

trouble arose over a stoker's refusing duty, and demanding to be put on shore. It is to be regretted that he had indulged in some of that fiery concoction procurable in such places, and known amongst sailors as 'chain-lightning.' This horrible stuff appears to madden rather than to intoxicate, and judging by his maniacal outpourings, he must have sampled a considerable quantity. He was so overbearing and desperate that he was put in irons and locked up in the sail-room.

The clink of the windlass in heaving up the anchor seemed to madden him afresh, as the gist of his threats was to the effect that he was determined not to proceed further in the ship, and if forced to, there would be wholesale murder, and not a single man's life was to be spared. In his madness he armed himself with an ice-pick and proceeded to demolish his prison door. Seeing that this must necessarily happen, for he was a powerful man, the captain directed Evans to take the two strongest men he could pick, and guard the door. Evans was always a peace-loving fellow, but, when there was fighting to do, he could be extremely useful and ugly.

'CHAIN-LIGHTNING' QUELLED

The stout door at length fell, in a more or less splintered wreck, out on to the deck. The infuriated man, with nostrils dilating and bloodshot eyes, and holding an ice-pick aloft, dared anyone to approach on peril of his life.

In a flash Evans leapt through the half-open door, and rapidly throwing his right arm round the madman's neck, squeezing it as though in a vice, he dragged the astonished wretch on to the deck, threw him flat, and delivered such physical punishment that in a very few moments all thoughts of murderous intentions had vanished.

The man spent that night shackled, with arms behind him, to a stanchion, and before noon next day, being stiff, sore, much bruised, and extremely hungry, he expressed the desire to be released, and promised that he would return to work. That man became one of the hardest workers in the ship, and was always most respectful and civil to all, especially to Evans.

After several days' sail the Line was approached, and as a considerable number of our crew had not in the course of their journeyings penetrated into the realms of the southern hemisphere, it was deemed imperative that they should be initiated into its mysteries by being presented to the Court of Neptune. The usual medical treatment and shaving operations were officially carried out, and the proceedings culminated in the recognised rough-and-tumble aquatic chaos of both victims and officials.

Among the more stubborn victims to undergo treatment was our friend Evans, who was captured in a lifeboat with a rope round his neck after an exciting chase by the constables.

With the advent of steam, however, these old customs of the sea, so religiously carried out in the old wind-jammer, have had their day, and are fast fading into oblivion.

THE AUTHOR

CHAPTER V

First taste of bad weather—Run across the Southern Ocean—Concerts and nigger minstrel entertainment—Sighting New Zealand—A beautiful dawn—Arrival at Lyttelton—Hospitality of New Zealanders—Preparing for the Antarctic—Sailing day—Humorous incident at farewell service—An impressive send off.

THE passage through the tropics was enjoyable, and it was not until rounding the Cape of Good Hope that any heavy weather was encountered.

Now, I had always regarded my chum as an exceedingly capable and efficient officer, but at the same time his sailing experience had been confined to naval training brigs, manned by a working crew, numerous enough to perform any manœuvre with rapidity.

As our crew comprised only five men in a watch, I hinted that he would have to make allowances for the proportionately

THE VOYAGES OF THE 'MORNING'

slower work, and not be tempted to carry sail too long with approaching bad weather. It was a friendly suggestion on my part, having experienced in sail the capabilities of a limited crew. The following incident will serve to illustrate how easily one may be taken unawares.

It occurred some days after rounding the Cape. The wind had freshened all day from the north, and in the evening, during my watch, it had veered to the N.W., and was blowing very strongly. The sky was overcast, and there was a spitting rain at times. The top-gallant sails were set, but all light stay-sails had been taken in. The sea was roaring and hissing up on the port quarter, and the little ship bowled along at nine knots. The wind, although so strong, was steady in force during the whole watch.

Evans relieved me at midnight, and expressed delight at the strong fair wind. We were both enthusiastic at the progress being made, and filled with that exhilarating feeling which prevails under such conditions. For what sensation can be compared to a ship driving before a

EVANS AND ENGLAND

A SUDDEN SQUALL

turbulent and seething ocean, with the wind, like an infuriated fiend, shrieking through the rigging ; the sails, discernible only against the scudding vault as dark oblong shapes, bellied out to their utmost limit ; the heaving vessel burying her lee in the foaming waters; the startling crack and creak of braces and sheets with the extra strain ; the gunlike bangs of the wash ports with the escape of the lee water ; in short, the animated wilderness of the environment ? Those who go down to the sea in ships appreciate these things, and can realise the majesty of the mighty ocean.

I had not been an hour below, after my watch, when a violent lurch of the ship startled me from my sleep and, quickly judging by the redoubled fury of the wind that a gale was upon us, I hurriedly put on a pair of slippers and rushed on deck in my pyjamas. The night was wild to behold ; the wind roared in its violence, a stinging storm of hail was beating the ocean into a boiling spume, and hard-driving spray was madly breaking over the entire ship, even up to the lower yards. The stout little craft was straining heavily before the elements, with lee-rail completely engulfed in the angry sea.

There was no sound or sight of human being, and I had quickly noticed that the top-gallant sails were still set. Groping my way aft as quickly as possible by clinging to the weather rail, I reached the poop, and, faintly through the fury of the night, I heard a familiar voice. 'Is that you, Jose?' it said; 'give me a pull with the weather top-gallant brace.'

It was pathetic. Here was poor Evans all on his own, struggling with practically the impossible under such conditions. But it was the only thing to do then. A sudden blinding squall had broken over the ship. He had fortunately seen its approach in time to place an extra man at the lee wheel, and in so doing had only narrowly averted broaching to. One man was on the look out, and the other two of his watch had been despatched to let fly both top-gallant halliards, to call all hands, and, if possible, to brace the fore top-gallant yard round to spill the wind out of the sail. Owing to the excessive violence of the squall, however,

FRIENDSHIP IN PYJAMAS

the sails, holding so much wind, would not permit the yards to come down until four or five men, after much struggling and straining, managed to point them to the wind.

Evans and I stuck to the main topgallant brace, and practically had the yard round before help arrived. He was exceptionally muscular, and I also apparently succeeded in doing herculean work, as it was only by almost superhuman efforts to retain circulation that I could possibly have withstood the rigours of such a night in a pair of cotton pyjamas!

The captain was, of course, soon on the scene. Anticipating a shift of wind, and the glass being very low, he ordered the mainsail and upper topsails to be taken in also. We worked away with a will, and before the watch was over, the ship was ' snugged down ' and prepared for the worst.

The strain was now over, and comparative peace reigned. At 4 A.M., as soon as he had handed over the watch, Evans and I went below, drenched to the skin, dirty and sore, especially about the hands and arms. Having clothed ourselves in comfort, we brewed some stimulating hot rum toddy, and completed the luxury with cigars. We sat thus for a considerable time, chatting over the strenuous evening's entertainment. Evans used often to relate this experience in after times, and elaborated on my assistance. 'That's what I call friendship,' he would say. 'Fancy a chap turning out in a howling gale in his pyjamas to give me a hand.' But it was, after all, my duty, and besides, it was a pleasure to be able to help my good friend, who had already done so much for me. After this night's experience I never needed to consider Evans's resources, which were quite equal to anything that might befall a ship even with a limited crew.

Gales of varied duration and intensity were more or less conspicuous along the Southern Ocean. The engines were seldom needed as the breezes blew the ship along a great deal faster, and economy of coal was thereby effected.

A fair amount of hard work, interspersed with a variety of entertainment, caused the days and weeks to pass merrily by.

'NIGGER MINSTRELSY'

When the weather was favourable, concerts were held, and to celebrate Evans's birthday (in October) he and I, aided by a young midshipman, inflicted a nigger minstrel entertainment on the company. The wardroom table was removed and chairs and camp-stools arranged to represent stalls and pit. The performers were attired in white mess jackets, dark trousers, and red cummerbunds, with the usual eccentric and extravagant stick-out collars and ties. Faces were, of course, blackened with the amateur make-up, burnt cork; oakum wigs adorned the heads; and the result was effectively ludicrous.

The items rendered were more or less topical, and Evans, being a good mimic, was especially happy in his humorous Dago stories. A stump speech on Geography gave opportunity to discourse nonsensically on the Antarctic and its uses, and a few humorous and chorus songs exhausted our repertoire. The audience seemed to enjoy the tomfoolery as much as we did ourselves, and a vote of appreciation was moved by the captain, who complimented our efforts

THE VOYAGES OF THE 'MORNING'

by remarking that he had paid money to listen to a very much inferior entertainment!

Possibly for our good, this flattering speech was at this point suddenly curtailed by the visit of a heavy squall, and the entertainment broke up unceremoniously with a wild exit of both audience and performers to take in and furl the top-gallant sails.

With time so fully occupied, there was no room for monotony or ennui, in spite of the fact that the outward voyage occupied four months. Early in November the *Morning*, true to her name, arrived at the first streak of dawn off Lyttelton Harbour, New Zealand. Here, after a period of three and a half months of eternal sea and sky, was the good land once more, and the breaking day displayed a beautiful and peaceful picture. The little ship, steaming through the placid water with yards squared to almost mathematical precision, sails snugly furled with an extra neat harbour stow, the burgee of the Royal Corinthian Yacht Club fluttering at the main, and the Blue Ensign at the peak, gradually approached the bold headlands of the harbour entrance.

NEW ZEALAND HOSPITALITY

On the one hand, along which we were steaming, were the hill-slopes of Banks Peninsula, indented by numerous picturesque bays, and on the other, in the grey distance across a vast expanse of low level plains, rose imposing snow-clad mountain ranges, partly wreathed in the morning mists, their roseate, sun-kissed peaks rising above.

Lyttelton, the picturesque town nestling on the slopes of the Port Hills, was soon reached, and in spite of the early hour, a welcome in the shape of hooting whistles, dipping flags, and much cheering was accorded the little ship's arrival. Thus the initial part of the *Morning's* programme was brought to a successful conclusion.

It would take many pages to attempt to describe the great kindness and hospitality extended to us by the people of Christchurch and Lyttelton, of every class. It was astonishing the interest displayed in the Expedition, and one and all seemed imbued with a similar enthusiasm for its success as we ourselves. The *Discovery* had, of course, materially paved the way twelve months before.

THE VOYAGES OF THE 'MORNING'

Time was limited, and a great deal of work had to be done finally to equip the vessel for her southern voyage. The evenings were mostly given up to social entertainment, which was liberally showered upon all members. Evans and I went everywhere together, and were known as the 'Doorvans Brothers,' until a friend with a waggish turn, suggested as an improvement, 'The 'Evanly Twins,' which was forthwith adopted!

It was remarkable what contrasts were experienced in evening entertainments; one evening we might be driven out in a carriage and pair to dine and stay the night at one of the most sumptuous and beautiful palatial residences, whilst the following evening there might be a hospitable welcome to coffee and cakes in the kitchen of a modest working man's family. We enjoyed these evenings, in whatever social sphere, because of the frank and genuine hospitality extended by all, both rich and poor, with whom we came in contact.

By the first week in December, the *Morning* was ready to proceed to the Ant-

A FAREWELL SERVICE

arctic in quest of the *Discovery*. The vessel had been docked, cleaned, and painted; rigging and gear carefully overhauled, and the hold laden to its utmost capacity with coal and stores. Every available locker and recess was likewise filled, and the large and interesting-looking mail for the *Discovery* occupied a room in itself.

Sailing day was perfect, ships were gay with streamers of bunting, and an immense crowd had assembled to express their last wishes of good cheer and God-speed. Just before departure, an impressive farewell service was held on deck by the Bishop of Christchurch. It was a simple and touching ceremony, but, even at the gravest moments, humour occasionally can be found lurking in some odd corner. Evans and I, sharing a hymn-book, were singing lustily, when one of the crew who, through misdirected kindness, had been given too much liquor ashore, and had been sleeping the effects off, appeared on deck somewhat stupefied. Seeing the group on deck, and evidently thinking that he was missing something, he edged himself between Evans and me. Respectfully, but unfortunately audibly, he asked: 'Ain't I going to 'ave me photo taken too?' This was a passing indiscretion; he was otherwise a very good man.

At length the moment of departure arrived, ropes were cast off, and the brave little vessel steamed slowly out of the harbour, amidst a perfect furore of cheering and handkerchief-waving. Whistles were blown from all directions, guns fired, and flags were flying gaily. On the tug which escorted the ship down the harbour were gathered all the good friends who had so generously assisted the Expedition in innumerable ways, and a band on board played appropriate and stirring tunes, until the final parting at the Heads, when, as a climax to an impressive send off, 'Auld Lang Syne' was solemnly played.

There was a sense of loneliness after all this excitement, and the whole-hearted response and affection of the many friends left behind had made a deep impression on us.

Gradually the land receded, and slowly the shadows of night crept over a peaceful

DEPARTURE FOR THE SOUTH

ocean. With thoughts of home and loved ones, and with high hopes for the future, our good ship gathered way, and through the lowering pall of night, she steered for the Frozen South.

CHAPTER VI

Southward ho!—The stormy Southern Ocean—First experience of ice—The midnight sun—Beauties of Antarctica—Christmas Day—New island discovered—Ship strikes an outlying rock—Captain's skilful seamanship—The ice pack—Seals—Exercise and sport—Dangerous gale on edge of pack.

THERE are few oceans so tempestuous as that globe-encircling expanse to the southward of the South Atlantic, Indian, and Pacific Oceans, usually known as the Southern Ocean. During all seasons stormy conditions prevail, and in spite of its being midsummer, the *Morning* encountered her full share of buffeting from the elements on her southern voyage. Being very heavily laden, the great seas broke continually over the ship, and one night during a gale one of the quarter boats was dragged out of its tackles and swept away.

Owing to these persistent westerly disturbances, the vessel was driven to leeward of the intended longitude for entering the ice

CHRISTMAS DAY

pack, and when the first drift ice and stray bergs were met, the ship was approaching the Antarctic Circle on the 180th degree.

This was Christmas Eve, 1902. The night was calm and wonderful. Here for the first time was perpetual sunshine, the sun just skimming the southern horizon at midnight. The peaceful ocean was dotted with innumerable masses of detached ice ; the oblique rays of the sun cast weird shadows across the jagged bergs, and reflected through their fissures and weather-worn caves the most superb tints of sapphire and azure. New and beautiful bird life now abounded, the pretty brown-backed and silver-grey petrel, and the graceful snowy petrel, which so perfectly matches the ice that when flying low its dark bill alone is discernible.

Christmas Day was even more perfect, the sky being cloudless. The calm sea, strewn with bergs of the most erratic shapes, and with denser drift ice, displayed in the intense sunshine a truly wonderful and dazzling effect. It was interesting, and indeed exciting at times, working the little ship through this icy archipelago.

THE VOYAGES OF THE 'MORNING'

The ward-room was decorated with gaudy-coloured sledge flags, and the numerous parcels labelled 'Not to be opened until Christmas Day' were duly laid out, and, after breakfast, investigated with all the customary eagerness and joyous abandon associated with Christmas.

Evans received four Christmas puddings, and in a parcel from his mother were calendars for himself, the captain, and me. There were also boxes of sweetmeats, including crystallised fruit, and a large and handsomely designed Christmas cake from New Zealand.

It was such a beautiful day that there was no such thing as worry or anxiety. We were a joyous party, and kept Christmas in the true British fashion. During the afternoon, Evans, who was on watch, added greater interest to the day by sighting an island, of which no record could be found in books, journals, or charts. With great pride, therefore, we claimed the discovery of an unknown island. In the distance, with so many bergs about, it might have been a large discoloured berg itself, but it was soon observed to be a typical Antarctic island, a portion being covered with the usual heavy ice-cap.

OBSERVATION HILL

PART OF SCOTT ISLAND AND HAGGITT'S PILLAR

A NEW ISLAND

Early in the evening the island was reached, and was estimated to be two miles in length, with a remarkable cone-shaped islet close off it. The ship being stopped, a boat was lowered and a party, after some difficulty, landed and officially claimed the island in the name of Great Britain. Three cheers were given by the shore party, which were responded to with enthusiasm from the ship. A record was left on the island stating the discoverers and date. The ship lay off the land for the night, and a keen discussion arose that evening as to its naming. Christmas Island was, of course, the most appropriate, but there were already several Christmas Islands in the world. It was eventually decided to call it Markham Island, after Sir Clements; but this was subsequently altered to Scott Island.

Next day the forenoon was devoted to making a rough survey, and to plotting the geographical position of our discovery. Here was an opportunity for our surveyor, Mulock, who kept us all busy taking sextant angles, altitudes, chronometer times, soundings, and speed registers. It was arranged that lines of soundings be taken from close in shore direct out to sea on the four cardinal points, to ascertain the receding depths.

These were carried out successfully from east to west, but, steaming about four knots along the land to get into position for the run out to the southward, the *Morning* very unfortunately struck on an outlying rock. This was a most embarrassing situation, especially at an unknown island in a desolate ocean, and we all admitted afterwards having enjoyed brighter moments.

'Full astern' was immediately rung down, and the feeble engines in response struggled bravely, but with no effect. In an instant all hands were available, and set about their work with vigour and alacrity. A slight swell caused the vessel to bump with heavy, sickening thuds on the rock, which appeared to be under the main mast. The ship shivered and shook with the continuous jarring, and, being wooden, yielded in a remarkable manner to the concussion. A steel vessel would simply have foundered. Masts and yards quivered like aspen leaves, boats being provisioned swung in the davits, and the whole scene was as animated as it was awesome.

STRIKING A ROCK

Our cool and collected commander lost not a minute. He was an able man, and well trained in the sailor's foremost qualification —resourcefulness—which now meant to think quickly, act promptly, and that rightly. A light breeze blew along the land. The order was given for all square sail to be set aback. The sails were set aback with wonderful rapidity, Evans and I competing as to who could get the sheets home first. The engines, working continuously, now acquired material acceleration by the leverage from the breeze against the spread of canvas. Just as a few crushed and splintered pieces of oak shot ominously from the bottom and fell with a splash on the surface of the water, sternway was perceptible, and after a few moments of suspense, the ship was boxed off the hidden danger, and slid into the gratefully deeper water. Yards were swung round willingly, and with the favourable breeze, the ship headed away from our treacherous discovery, and sailed for the southern pack. The captain calmly picked up his camera and took a final snap-shot of the island, ' to show there was no ill-feeling,' he said, and also remarked that the line of soundings to the southward could await a more favourable opportunity!

In a few hours' time the dense ice pack was encountered, and varied progress was made through the night; but on the following day the ship was brought to a standstill in an ice-covered sea. As far as eye could discern from the crow's-nest, there were blocks of every conceivable shape—large flat floes, and bergs packed together in one impenetrable mass. There appeared to be no way onwards in any direction, nor, indeed, was there any trace to show how the ship had arrived thus far, all wake-tracks being completely obliterated.

This great belt of ice, which has to be penetrated for one to two hundred miles before gaining the comparatively open sea once more, is the result of the winter's ice on the outskirts of the Antarctic shores, which drifts to the north until, reaching the warmer latitudes, it ultimately melts away. Owing to this drift, the entire sea of ice is continually on the move, and even if a ship is so beset at one moment that it

IN THE PACK

seems apparently hopeless ever to get free again, in the course of a few hours, perhaps, the aspect may be so altered that long water 'lanes' open out in many directions through which the ship can work for miles at times, until she becomes temporarily beset again.

And so it goes on for days on end, with the ship ever straining and quivering, similar in effect to a traction engine struggling over a stony roadway, the roar along the sides sounding from below like incessant distant rumbling thunder, with occasional dull thuds when an ice block, smashed down by the ramming bow, bumps along under the ship before refloating.

The pack varies in intensity and area according to close or open seasons. In the vicinity of land, the pack is generally very heavy, and owing to this obstruction, it is banked up, and much pressure arises. This is a source of grave danger, as a vessel runs the risk of being squeezed, and perhaps utterly crushed. Well out to sea, however, navigating in the ice pack is interesting, though at all times it sorely taxes the patience.

THE VOYAGES OF THE 'MORNING'

When beset, we always embraced the opportunity to take exercise on the large floes; ski-running, foot races, or football being always popular forms of recreation. Watering ship was done from time to time, and there was always a standing occupation of capturing seals and penguins. The crab-eater, or white seal, and the Weddell seal abound in the pack, and at rare intervals a sea-leopard is secured; the latter are formidable creatures, measuring as much as twelve to thirteen feet in length, with a girth of six feet, and demand much energy on tackles to hoist on board. The penguins, however, always afford much amusement by their quaint antics, and with their black backs terminating in the funny little tails, spotlessly white breasts, and flippers akimbo, humorously caricature little men in evening dress suits! A fit of the blues could readily be dispelled, I should imagine, by just watching these unconsciously humorous creatures.

The New Year was brought in in the authorised manner by blowing the whistle, ringing the bell, and general noise. Evans relieved me as usual at midnight, and after

AN EARLY NEW YEAR

firing off a lumbering rifle known as 'The Arquebus,' with a report resounding enough even to cause the drowsy old seals to move their heads slightly, we wished each other the usual good things, apropos of New Year, and talked of the dear ones at home, knowing full well how they too would be thinking of us at this time. As the ship was practically on the one hundred and eightieth degree, we contended that we were bringing in the New Year before any other people in the world!

After four days of battling, having worked into the middle of the pack, a gale set in from the eastward, but only by the whistling of the wind through the rigging could one realise that such was the case. The ship lay motionless among the heavy floes, as quietly as at a wharf, and it was difficult to imagine that the ocean was, at this point, two thousand fathoms deep. No sea can rise, nor swell assert itself, owing to the presence of such masses of densely-packed ice.

Progress to the south was made as opportunity offered, and after three more days,

THE VOYAGES OF THE 'MORNING'

the pack was noticeably looser. The gale, however, had blown persistently, and was even increasing in violence. It was clear that the southern edge of the pack was being approached, and the experience of a gale there was not cordially welcomed.

The heavy floes and hummocked blocks, so beautiful and peaceful-looking only a few days before, now assumed a totally different aspect. The sea was gradually rising, a swell perceptible, and the ship bumping with increasing thuds against these rolling spur-projecting dangers. The full force of the gale soon threatened to overwhelm the stout but weak-powered little ship.

In a very short time the whole ocean was a seething mass of storm-tossed ice derelicts, and the fury of the wind did not permit of any canvas being set. The engines were kept going for every pound they were worth—which power, however, hardly accomplished steerage way. It was a tight corner, but it had to be faced, and with as much skill as possible. Among such numerous heaving dangers, it was impossible to avoid occasional collision, and there would be a

A STORM IN THE PACK

truly sickening shock, when a great berg, sobbing and seething as it rolled, with the water hissing and belching out of its fissures, would be dashed from the wave's very crest with a terrific crash against the almost helpless vessel.

However, by dint of constant vigilance in endeavouring to weather the more formidable masses, and by giving way to others, though anxious in mind, weary in body, eyelids drooping, and faces cut by icy blasts, we succeeded in conning our little ship safely through a seeming eternity of four awful days and nights.

CHAPTER VII

> First glimpse of polar continent—Cape Adare—Plans of Relief Expedition—The humorous penguins—
> —' Southern Cross ' Expedition hut—First *Discovery* record found—Along the coast of South Victoria Land—Immense tabular bergs—Possession Islands—Coulman Island—Impenetrable ice pack—Futile efforts to reach Wood Bay—Compass error—Strange ski impressions—Desperate attempt to reach Franklin Island — Providential escape from insetting pack.

THE tempest abating, nature, by way of contrast from her recent raging, smiled peacefully, and once more bathed the world in sunshine. Upon the smooth waters, the broken floes and weather-beaten bergs, floating listlessly, sparkled in the sunlight. Wide waterways were now revealed in many directions—black tracks against the dazzling snowy glare. The prevailing stillness was only disturbed by the movement of the ship or by an occasional penguin's caw.

CAPE ADARE

Steaming ever southward, the ship charged at times through loose areas of ice, ploughing her way onwards with little reverse. In the far distance to the South, loomed, gradually, the first glimpse of the polar continent, and by midnight the lofty, ice-covered Admiralty Range of mountains stood boldly silhouetted against the golden blaze of the southern horizon. Cape Adare slopes to the north from this range, and forms the eastern side of Robertson Bay.

After one full day's steaming, without unusual incident, the ship arrived at this point. On the flat, stony foreshore thousands of penguins could be seen, and, on closer observation, a dwelling hut, standing in the middle of the rookery. The 'Southern Cross' Expedition, of which our captain was a member, had, some years previously, erected this hut, in which a winter was spent.

The boat being lowered, a party, after negotiating a way through some stranded ice, landed, and with much interest wended its way through the rookery to the hut. They were greeted with a perfect bedlam of cawing penguins, making a really deafening and bewildering uproar. The breeding season was in full swing, and there was much animation in the settlement, especially after the intrusion of strangers. The young penguins, little grey, fluffy blobs, squeaking in their terror, excitedly sought the parents' protection; whilst the adult birds in turn, with head comb raised angrily, gave vent to a persistent shrieking caw, and, in perfect paroxysms of uncontrolled temper, lashed out viciously at the men's legs with their hard little flippers. What quaint angry little things they are! When a desperately irate penguin rushed at one with eager trembling flippers, one had simply to put a foot out, and the foolish creature, mad with rage, would hurl itself blindly at the point of the boot. The momentum, thus unexpectedly checked, would cause the unfortunate bird to describe an undignified backward somersault! Struggling to its feet again, and for a moment apparently stupefied, it would reel as though tipsy; then, with redoubled energy, another blind charge would be made at the boot, with the same upsetting result! These pen-

THE FIRST CLUE

guins seem to lose control of themselves through anger, actuated, no doubt, by fear; but whatever they do, they display wonderful determination, and always appear to be in deadly earnest.

The outer door of the hut being securely nailed, was, we considered, an ominous sign, but by means of a crowbar it was levered open, and on the top of an old cask standing in the porch, was a red painted tin cylinder with the name *Discovery* roughly painted on it in white letters. This first clue in our search for the *Discovery* was opened with much eagerness in the living-room of the hut. It proved to be the first record left by that ship after leaving New Zealand, and was placed in the hut exactly twelve months before our visit. The only information gleaned was the fact that the *Discovery* had reached Cape Adare safely, having negotiated the ice pack in six days; all on board were in excellent health and spirits, and the ship was proceeding on her voyage south the same day. A number of letters from staff and crew were also found in the cylinder.

Before leaving England, it was arranged

for the *Discovery* to leave records at certain points along the coast, and the *Morning's* duty, following twelve months afterwards, was, if possible, to find these records, and thereby locate the *Discovery*. This scheme, however, could not be otherwise than crude, by reason of the limited knowledge of the various points of vantage. Cape Adare was a certain mark, because of its hut, but other points and islands, being of considerable area, with undefined marks, rendered a search for so small an object about on a par with looking for a needle in a haystack. Still, the *Morning* had thus far penetrated the main ice pack, reached Cape Adare, and had found, in spite of its meagreness, the first clue of her search.

The hut at Cape Adare is interesting, being a strong, well-built protection. Articles of clothing and stores lay about the living-room, and everything was in such good preservation that it was hard to realise that the place had been for several years uninhabited. After a few pleasant and interesting hours at Robertson Bay, the ship proceeded to sea, and rounding Cape Adare, steered to the south-east.

All day long the most imposing and

TABULAR BERGS

stupendous tabular bergs were passed. These bergs of barrier formation are peculiar to the Antarctic, and are so vast as to be termed ice islands. The length of some observed was estimated at four to five miles, attaining a height of one hundred to one hundred and fifty feet. The *Morning*, steaming five knots, would often take an hour to pass one of these immense floating masses, and the top surfaces could just be seen from the crow's-nest at the mast-head.

A group of small islands, known as the Possession Islands, was next visited, and a landing effected on the largest island, where the most conspicuous mark was also a penguin rookery. A thorough search of the beach was made, but no trace of a record was found.

Coulman Island, a large round-backed, ice-capped island, was next steered for. The island lies close off the mainland, and a vast quantity of ice was densely packed all round it for miles. For four days repeated efforts were made to reach the island from every available quarter, but it was found impossible to effect a landing, the nearest approach being about fifteen miles.

THE VOYAGES OF THE 'MORNING'

Seeing that it was futile to think of reaching the island, and as valuable time and coal were being expended, an attempt was made to get into Wood Bay, a pretentious inlet in the mainland some miles south of Coulman Island.

Wood Bay was considered a likely place for the *Discovery* to winter, being apparently free from ice pressure, and having the advantage of closer proximity to the south magnetic pole than any then known haven. The magnetic pole, or southern extremity of the earth's magnetism, being approximately two hundred miles inland, affected the *Morning's* compass to the extent of 147° deflection to the eastward, or, roughly speaking, if a true south course was to be steered, the compass pointed N.E. by N. This fact should be of unusual interest to nautical men.

A determined effort was made to force the ship through the outlying pack, and into the bay, but it was a forlorn hope. An absolutely impenetrable ice pack extending twenty to thirty miles out to sea, confronted the ship, and for two days, skill, energy, and coal were expended in vain. Discussion arose as to the feasibility of dispatching a party to sledge

AN ANTARCTIC PILLAR-BOX

DELAYED FOR A WEEK

[See p. 89

STRANGE SKI-MARKS

across the ice, but owing to the vagaries of ice packs in opening out unexpectedly, or drifting, the risk entailed was considered too great, and the idea therefore abandoned. This was now the third disappointment in our efforts to obtain tidings to direct our difficult search voyage, and it was discouraging and perplexing in the extreme.

Needless to say, everyone was at high tension, and keenly on the alert for any possible sign to identify the *Discovery's* whereabouts. One day, off Wood Bay, a large floe was passed with two deep ruts, equi-distant apart, over the entire length of the floe. This exactly coincided with the impression made by skis on the snow surface, and for a considerable time it was a matter of exciting speculation. However, it was discovered that penguins, although seldom employing this method, do occasionally travel along the snow on their breasts, propelling themselves with their feet, and the result is a very close resemblance to the impression made by skis.

Visible some miles to the southward, and well off shore, was Franklin Island—a long, narrow island, partly covered with the usual ice-cap—which, from its central situation and

THE VOYAGES OF THE 'MORNING'

comparatively easy access, was the most likely place for a record to be left.

All hopes of reaching Wood Bay being abandoned, the ship was forced well out to the eastward to clear the heavier land pack, and Franklin Island steered for. It was soon discovered, however, that the pack surrounded even this island, and extended several miles out to sea. The time had now arrived, though, for drastic steps to be taken to gain some information. A desperate attempt was, therefore, made to reach Franklin Island.

The pack, composed of exceedingly heavy, hummocky floes, rendered the forced passage doubly trying, and the ship, charging against such obstacles at full speed would receive frequent severe shocks, and be brought up all standing, quivering with the concussion. Many hard hours were spent in this determined struggle, and even when about one mile distant from the island, the prospects of success were doubtful, the ice being heavier and closer packed, and, worse still, a lane of open water, perhaps half a mile wide, lay between the pack and the island. If the ice had extended to the shore, one could safely have walked across

TO FRANKLIN ISLAND

it and landed. The smallest ship's boat was too cumbersome to drag across the hummocked surface, otherwise this would have solved the problem.

It was a tantalising situation, but as the ship was now in such a position that to proceed or retreat was equally arduous, it was decided, after an hour's adjournment for lunch, to adopt the former move, and endeavour to push through the intervening half-mile. This was no easy matter, as may be imagined, and the ship suffered perhaps the most excessive buffeting she was ever to experience. By going astern, the rudder would bring the ship up against the ice, then ' full ahead ' for a few seconds, and the sheathed bows would crash into a veritable wall, bringing everything to a standstill. Then astern again, and again ahead, and so on without cessation.

It was bold and desperate work, but it was crowned with success, for after two hours of hard going, we had the satisfaction of seeing the last floe yield, and the ship gather way into the narrow water-lane. Ice-anchors were placed, and the ship moored against the inner side of the last troublesome floe.

THE VOYAGES OF THE 'MORNING'

A boat was soon away, and the party landed on the long sloping stony beach, occupied as usual by thousands of penguins.

The ship had been in this position for about an hour when a very unexpected and unwelcome change occurred. A dark, threatening cloud bank rose from the south, the whole sky soon became overcast, and the general outlook appeared extremely gloomy. As a blizzard might evolve suddenly, signals were made to recall the shore party, who were, however, so engrossed in their search that the signals were unobserved. The whistle was blown and detonators fired, but to no purpose. Then the whole ice pack, for some unaccountable reason, commenced moving gradually and eerily towards the land, and it seemed inevitable that the ship would very soon be set broadside on to the rocky shore. Every effort was made to attract the shore party, who, fortunately, noticing the ship's predicament, hastened back on board. They were only just in time; indeed, the ship was under way before the boat was hooked on for hoisting. A strong breeze was now blowing, and the driving snow threatened to obscure everything. It was

A NARROW ESCAPE

imprudent to attempt to force a way through an incoming ice-field, and the only chance of escape, risky though it appeared, was to steam as quickly as possible along the edge of the island, before the ice banked up against it. It was a touch and go, and the odds were against the ship, which was forced to skirt the shore ever closer and closer as the ice encroached on the limited water lane. Outlying reefs or sunken dangers, if there were any, could not be reckoned with in the dire predicament, and the points of the island were narrowly skimmed, as there was only bare room then for the ship to pass.

After running this anxious gauntlet for the entire length of the island, to our intense relief the pack became looser, and soon several exits in the shape of narrow water lanes providentially presented themselves. In a few hours' time the ship was pushed through this looser pack, and with much delight the open sea was once more welcomed. The hardest feature of this trying ordeal was that, in spite of its exertions, anxieties, and risks, no record was found.

CHAPTER VIII

Difficulties attending our undertaking—Cape Crozier and Mount Terror—Depots to be established if no record found—Party at Cape Crozier—Discovery located—Delayed a week in heavy pack—Mount Erebus—The deadly stillness—Beaufort Island—A sailor's venture—Subsequent humorous sequel—Exciting steam up MacMurdo Sound—Sighting Discovery's masts.

It may now be understood how very difficult and arduous a matter it is to track a vessel in these more or less unknown regions. So much has to be reckoned with that the most carefully prepared schemes, combined with the utmost resourcefulness and ingenuity, are only too frequently doomed to disappointment and failure.

When the excitement of the hair-breadth escape from Franklin Island had subsided, the résumé of the search voyage so far had to be summed up as disappointing and unsatisfactory. The places visited had produced no

CAPE CROZIER

information, and the inaccessible points might, or might not, possess some clue. All that was possible had been done to obtain tidings, and the result was—perplexity.

The only remaining course now was to investigate the last definite point, which was, as far as we knew, the most southerly land known.

Having been extricated from all pack, the *Morning* was steered south again in the open waters of the Ross Sea, passing a few enormous bergs, and occasional drift ice, but nothing in the nature of a pack.

By nightfall, after steaming all day from Franklin Island, the dark outline of Cape Crozier could be seen ahead, and above, through breaks in the overhanging clouds, glimpses of Mount Terror were disclosed. Cape Crozier forms the termination of the eastern slopes of Mount Terror, immediately behind being the commencement of the Great Ice Barrier, which trends for miles to the eastward.

If no record were found at Cape Crozier the *Morning's* programme would be materially affected. In view of this eventuality, instructions had been issued for the ship to follow along the Barrier to the eastward as far as a certain indentation in its face known as the Bay of Whales ; and if no record were discovered even there, a depot of fuel and provisions must be established. The ship would then retrace her tracks and establish large depots of fuel, provisions, and clothing at all the points visited on the outward voyage, namely, Cape Crozier, Franklin Island, Wood Bay, Coulman Island, Possession Island, and Cape Adare ; on the completion of which, as the navigable season would be over, the ship must forthwith return to New Zealand.

To perform this depot-laying scheme would necessarily entail much solid work, and, judging by the recent experience of the difficulties attending some of the points, not a little risk. Cape Crozier was therefore approached with longing hopes for deferred information.

On drawing up to the land, the unfolding midnight scene was as impressive as it was beautiful. The sea was smooth, the air still, and the scene was tinted a purply grey. The heavy overhanging clouds gradually lifting revealed Mount Terror in its magnificence,

A RECORD FOUND

rising abruptly from the cape in a towering ice-clad cone of eight thousand feet, with hummocked, extinct volcanic craters along its steep slopes. Behind the cape rose the Great Ice Barrier, whose miraged surface scintillated dazzlingly in the sun's slanting rays. A reddish coloured slope from the cape indicated a penguin rookery, the prearranged mark for the Cape Crozier record.

A boat was soon away from the ship, and the eager party, on landing, scattered in search in all directions over the rookery. The ship cruised a short distance from the shore with an equally interested group, every movement of the shore party being observed through telescopes and binoculars, and to the men anxiously awaiting this almost vital result the search seemed interminable.

As a matter of fact the shore party, after the first hasty search, were becoming somewhat anxious themselves, when the captain, sighting a post standing well up in the rookery, raised an excited yell, and rushing wildly through the rookery, oblivious of the abject terror created amongst the unfortunate penguins, reached the post first, followed rapidly

THE VOYAGES OF THE 'MORNING'

by an excited and eager party. This smart movement was, of course, noticed from the ship, and a lively discussion arose as to its meaning.

In a few minutes a semaphore signal was made from the shore, and, almost breathless with suppressed excitement, this grateful and thrilling message was spelt out :—

'Found three records. *Discovery* in McMurdo Bay. Latest news only three months old. All well.'

It is not hard to imagine the feelings of the little ship's company. After all the reverses and disappointments of the past weeks, with their accompanying anxieties and doubts, finding such news at this final point in the search was altogether delightful. Lusty cheers were exchanged in an outburst of pride and thankfulness. The search party soon returned to the ship, and keen indeed were those who gathered round the ward-room table to learn the contents of these historic and interesting documents.

The first record, twelve months old, was left at Cape Crozier when the *Discovery* was on her way to the eastward. It stated

CAPTAIN COLBECK READING 'DISCOVERY' RECORDS FOUND AT
CAPE CROZIER
(Lieut. Mulock on right)

NEWS OF THE 'DISCOVERY'

that, after leaving Cape Adare, twelve months before, the ice-pack was too dense to effect a landing at the Possession Islands, but records were left at Coulman Island and Wood Bay. Franklin Island had also to be omitted owing to excessive pack ice. It was interesting to note, by comparing the two seasons, the difference in the ice-distribution along the coast, the previous season, as can be seen, being totally different.

The second record, eleven months old, was left on the *Discovery's* return, and contained an interesting account of the journey to the eastward along the Great Ice Barrier, when, after several days of steaming and penetrating some heavy ice-pack, a new land, named King Edward VII Land, was discovered. As the ice showed signs of much pressure, and as no inlet could be found for wintering, this inhospitable looking locality was left and the ship returned to South Victoria Land.

The third, and most interesting record to us, was left by a sledge-party the following spring, who, after leaving the *Discovery's* winter quarters, had found a route over the Barrier to Cape Crozier. The record stated that, after leaving Cape Crozier, the *Discovery* steamed along the Mount Terror shore to its western extremity, passing between this termination, known as Cape Bird, and Beaufort Island, a few miles to the northward. Proceeding still to the westward, across the entrance of McMurdo Bay (as it was then termed), a suitable winter harbour on the mainland was discovered, and named Granite Harbour, its only disadvantage being its exposure to the eastward. However, in the event of nothing better offering, this harbour could be availed of.

As McMurdo Bay appeared comparatively ice-free, the *Discovery* proceeded due south, and after steaming about forty miles up this inlet, the mountainous ranges of the mainland to the westward, and Mounts Terror and Erebus to the eastward, a small well-sheltered bay was discovered at the extremity of a narrow neck of land sloping to the south-west from Mount Erebus. To the south of this, again, a continuation of the Ice Barrier was found to cross the

SAILING DIRECTIONS

bay, and this revealed the surprising fact that this inlet, hitherto considered a bay, was in reality a strait, and Mounts Erebus and Terror formed a large island. The strait was subsequently called McMurdo Sound. Attaining such a high latitude as 77° 48′ south in the ship was eminently satisfactory, and the small bay at the termination of the Erebus slope was utilised by the *Discovery* as the winter quarters. ' A comfortable winter was passed,' the record added, ' the only real inconvenience being the persistent and severe blizzards from the south-east.'

With the advent of spring sledging was commenced, and one of the first journeys undertaken was along the Barrier surface, following the southern shores of the Erebus and Terror Island, and a way discovered to the eastern extreme of the land, terminating in Cape Crozier, where this last record was left.

The ' sailing directions ' in the record recommended following the *Discovery's* route from Cape Crozier, along the Terror shore, between Cape Bird and Beaufort Island, and up McMurdo Sound. There was also a rough chart of the sound, showing the more prominent and conspicuous landmarks, and a sketch of the aspect of the land and winter quarters from a distance of ten miles.

The information and instructions in this record were naturally read with the greatest interest, and, as the distance to the *Discovery* could not exceed ninety miles, hopes ran high that before another day had passed the long-looked-for meeting of the ships would eventuate.

The ship now headed to the westward, and, with extra willing hands, there was no difficulty in maintaining a full head of steam. Cape Bird and Beaufort Island were already visible in the grey distance, and a course was set to pass between these points.

The *Morning* had only been steaming for about two hours from Cape Crozier, however, when an ice-pack was observed ahead, into which she ultimately ran. For a few hours varied progress was made on a more or less westerly course, but the pack getting heavier and more compressed eventually brought the ship to a full stop. As far as could be seen from

A FINE-WEATHER CHECK

the crow's-nest was an endless pack, extending even to the northern horizon. This was a delay not allowed for in our calculations, and it now seemed that the ninety miles to the *Discovery* would not be accomplished in one day.

One day ? Why, this day passed, and the next, and the next, and the progress made is not worth recording. It was extremely disappointing, and quite evident that more pack remained this season than the previous one, in which the *Discovery* had sailed unchecked over this very track.

The weather was very fine during those days, which perhaps accounted for the sluggishness of the pack. The scene, though growing exceedingly monotonous, was nevertheless striking. Ahead were Cape Bird and Beaufort Island, between which points, and in the dim distance, the mountainous coastline of the mainland could be seen. To the south, now opened clear of Mount Terror, loomed the imposing ice-capped Mount Erebus, for ever emitting steam from its crater, thirteen thousand feet above sea level. The days were sunny, with variable breezes, but at night, the world seemed to stand still. The sun, passing low behind the darkened mountains, caused the diffused light to cast a weird gloom over everything, and the deadly silence, intense enough almost to be felt, produced indeed an eerie effect. Standing in the crow's nest, the pervading stillness was oppressive, and a ' chink ' of the wheel chains even would sound startling. Occasionally, miles distant, the sharp reports of breaking ice resounded alarmingly, like rifle cracks, showing that at any rate Nature toils ceaselessly though the world seems dead.

Although small progress was made with the ship as far as human skill was concerned, yet it was plainly visible that a ' set ' of some sort was forcing her along to the westward. Nature again assisted and saved much anxiety. By the fifth day the ship was but a mile or two from the channel entrance, and remained fast in very heavy pack, the current seemingly weaker.

One of the sailors, embracing this opportunity, jumped from floe to floe, and succeeded in landing on Beaufort Island, about one mile away. It was a risky venture, as, in the event

'NAME THIS CHILD'

of the pack's loosening, it might have been impossible to get to him. He returned safely, however, and received a reprimand for his pains. By digressing for a few moments the sequel to this incident may be related.

This sailor lived in New Zealand, and, one day after our return to Lyttelton, he confidentially informed Evans and me that his wife had recently presented him with a daughter, and he would be grateful if we could advise a name for the child, and would also esteem it a great favour if one of us would be godfather. We were then in Evans's cabin, the table and walls of which were decorated with photographs, including those of some pretty girls. The sailor was invited into the cabin and asked to choose the face he liked best among the photographs. Selecting one, Evans informed him that his choice was a girl named Myrtle, and he could, therefore, do no better than to suggest that name for his small daughter. For a second name Evans strongly emphasised the appropriateness of Beaufort, being, as he explained, not only a 'high-class' name, but, as the child's father was the only man, alive or dead, who had ever landed on the island of that name—and in so doing had incurred a severe censure!—it was only fitting that the name be adopted in his family, even if only as a reminder of his indiscretion. The child was therefore christened 'Myrtle Beaufort,' and Evans stood as godfather.

But to continue the narrative. During the afternoon the current became noticeably stronger, and the ship, together with the pack, was set rapidly, by this unseen power, through the narrows between Cape Bird and Beaufort Island. As soon as the channel was passed the pack loosened, and from the crow's-nest came the welcome report that open water was visible to the south. The engines were soon going their hardest, and in a very short while the pack was cleared and the ship, heading to the south, at last steamed into McMurdo Sound.

As far as could be seen the sound was apparently ice-free, and this, after six days' imprisonment, was a delightful change. Coal was piled into the furnaces, and the little ship, steaming a possible seven knots, trembled with the propeller's throb. The night was calm,

ENTERING THE SOUND

and lowering clouds hung over the higher land. To the westward the mainland of mountainous ranges revealed at intervals their bare granite sides of reddish-grey colouring and glacier-filled valleys. The whole of the eastern shore was bounded by the majestic Erebus, whose smooth slopes could be seen ever higher and higher through the overhanging clouds, with an occasional glimpse of the lofty steaming crater itself. To the southwest mountain peaks stood in bold relief against the midnight sky, and the southern horizon, over the Barrier, was veiled in the softest purple haze.

But greatest attention was directed to the southern slopes of Erebus, where point after point along its shores was keenly picked out, and verified on the rough little chart.

A prominent cape was passed, then some conspicuous dark islets, and a peculiar glacier tongue sloping well off shore ; next a remarkable landmark, Castle Rock, protruding darkly from the narrow ice ridge off the south-western slope of Erebus. The excitement was now intense, the sighting of the *Discovery* being momentarily expected. Every

THE VOYAGES OF THE 'MORNING'

available telescope and glass was brought to bear on the slowly unfolding scene.

The aspect now practically coincided with the sketch found in the record, and, by carefully following along the slope from Castle Rock, we noted the ridge terminating in a low, icy projection, named on the chart Hut Point! The *Discovery* must surely be there!

An extra keen sailor, who had been for six hours in the crow's-nest, suddenly began cheering, and pointed excitedly ahead. Through the most powerful telescope on the bridge, and with eyes straining, the trucks of two of the *Discovery's* masts were just discernible immediately behind Hut Point! The whole ship's company rushed eagerly on to the bridge, glasses were directed at this long-looked-for sight, and through sheer excitement and thankfulness a spontaneous stirring cheer rent the night air.

But if there were thoughts that the ship would steam close to the *Discovery*, they were doomed to disappointment; for in a few minutes an extensive ice-field was seen stretching across the entire width of the

FOUND AT LAST

sound, which soon brought the ship up at a distance of, approximately, ten miles from the long-looked-for goal!

Still, the arduous search was over: the *Discovery* had been successfully tracked in seven weeks after leaving New Zealand; and at midnight, through the quivering mirage, as a proof that we also were observed, a large Union Jack fluttered out a welcome from the staff at Hut Point.

CHAPTER IX

Ten miles of field ice—Experiments with ice-saws—First visitors—Return of sledge parties—A curious sight—Return of Captain Scott's southern party—First visit to the Discovery—The winter harbour—Meeting Scott, Wilson, and Shackleton—Banquet!—Captain Scott's speech—Astonishing hunger after sledging—An incident both humorous and pathetic.

PERHAPS only those who have experienced weeks on end in an unknown and uninhabited region can thoroughly appreciate the peculiar sensation created by gazing intently once more upon some sign of civilisation. Although quite anticipated, yet the actual sighting of the *Discovery's* masts, such mere specks in this eternal desolation, was both fascinating and impressive.

Much eagerness was displayed to sledge the mail promptly over to the *Discovery*, but the ice in the immediate vicinity appeared too unsafe to risk the transport of such valuable stuff.

ICE SAWS

Early the next morning tripods were rigged on the ice ahead of the ship, and ice-saws were tried with a view to cutting a way through to the solid field about two miles in. The ice was, however, in such a honeycombed and rotting condition that very soon the sawing party got into a sorry plight by frequently slipping through the slushy surface up to their knees, and, at times, to their waists. Finally, as the tripod itself threatened to perform the disappearing act, sawing operations were hurriedly abandoned, and, with difficulty, both men and gear were hauled on board.

The ice, although in such a rotting condition, was, nevertheless, so slushy and sticky as to check effectively all efforts to push the ship through it. Two miles in, it was several feet thick, and apparently continued solid to Hut Point, or further. It was decided, therefore, that patience was the only plan to adopt for the present : the rotten ice would soon break away, and the ship could then be secured to the solid field.

A constant watch was, of course, kept in the direction of Hut Point, and by noon came the welcome report that a sledge party had rounded the point, and was heading for the *Morning*.

Ten miles is a long way in these regions, but to the uninitiated and inexperienced this party seemed to travel all too slowly. About 3 P.M. a halt was observed, and two tents pitched—for afternoon tea, as we learned later! Re-packing the sledges again, every movement of our visitors was keenly watched, and gradually and steadily they approached. About ten men were counted, dragging two sledges, and the few dogs accompanying them seemed to enjoy their freedom in scampering for miles out of the way at times to yap at and torment any sleepy old seals unfortunate enough to be in the vicinity.

A signal was made from the ship warning the party of the rotten ice area, and although the message was received, their progress was little retarded. A leader, testing the strength of the ice with his ski staff, piloted the others in a most skilful manner to within a few yards of the ship.

Then a great cheering burst forth and much animation prevailed. By means of

'EVERY MOVEMENT WAS KEENLY WATCHED'

APPROACH OF OUR FIRST VISITORS

A MEETING EFFECTED

ropes and planks thrown out, these intrepid weather-beaten explorers were expeditiously hauled on board. As soon as the last man reached the deck, another spontaneous cheer rang out, and with the warm-hearted handshakes, the cheering, and the yapping of the dogs, the incident could not have been more thrilling.

Here, then, were the men we had come out to assist—men hardy, and tanned, and sturdy of frame. Officers, scientists, and seamen, each assisting the other in the one common cause—men of simple bravery, who could modestly and unaffectedly tell of terrible hardships, privations, and hair-breadth escapes, enough to make one's blood tingle with pride.

This party, under Armitage (an old *Worcester* boy), had only returned four days before from the pioneer journey up the western mountains, occupying two months. All other members of the ship were still away on various sledge journeys.

Captain Scott and Dr. Wilson—whose subsequent tragic and heroic deaths, after reaching the South Pole during their later

THE VOYAGES OF THE 'MORNING'

expedition, so completely shocked the civilised world—were away, in company with Lieut. Shackleton (now Sir Ernest Shackleton), on the first southern journey ever attempted. They had set out early in November for the south, Captain Scott expecting to return during the first week in February. If, however, through unforeseen circumstances, they were delayed, and had not returned by the first week in March, the hut (on Hut Point) was to be well provisioned, a search party left at the base, and the *Discovery*, with the relief ship, was to return forthwith to New Zealand.

During the next few days the rotten ice broke away, and the ship was secured to the main field.

Sledge parties were now returning, each with its fund of thrilling adventures and discoveries, and in turn they also visited the *Morning*, to welcome fresh faces, to engage in new conversations, and to learn the news of the outside world. A great demonstration always greeted the return of a sledge party. When sighted, parties would rush out to meet them, and amidst cheers and con-

KILLER WHALES

gratulations, they would harness in, and assist to drag the sledges home.

By the last day in January all parties were back at the base, with the exception of the southern party. A constant lookout, both night and day, was kept to the southward over the great wastes of the Ice Barrier. Fresh provisions and clothing were sledged over to the *Discovery* from time to time, the New Zealand mutton and potatoes being especially relished. Since an unfortunate outbreak of scurvy during the winter the entire company had lived on seal meat, discarding all canned meats as the seeming cause of this dreaded disease.

One evening a very extraordinary thing was pointed out by Evans, who had been scanning the southern horizon with a telescope. On the snow surface, about two or three miles south, were two conspicuous dark objects with rounded tops, similar in shape to large torpedoes. No one could hazard a guess as to what they were. Even if distorted by mirage they could hardly be men or seals, because the height of these strange shapes must have been twelve or thirteen feet, and they were also of bulky circumference. They were seen to be moving continuously from side to side. Through the most powerful telescope these weird-looking things proved to be the heads of two killer whales! Just in this vicinity were two or three seal holes which we had previously noticed when sledging, and the whales had presumably found these beneficial to obtain air, or more likely still, they may have been on a seal quest. They remained swaying monotonously for a long time, and Evans made a sketch of this unique sight, which I daresay he still has in an old diary. It is known, I believe, that killers do attack seals; at any rate it has been proved that seals have an enemy of some sort in the water. We never trespassed near these seal holes again, in case a sportive killer, seeing the dark reflection through the ice, might, in all innocence, mistake one of us for a seal, with a result not at all pleasing to think of.

While we were at breakfast on the morning of February 3 a report was sent below that the *Discovery's* masts were dressed with flags. A few of that ship's members being on the

RETURN OF SCOTT

Morning at the time, immediately assumed that the cause for this display must be the possible return of the southern party.

Hastily finishing the meal, all was soon animation and bustle. A couple of sledges were loaded, wind-suits and the usual sledging apparel donned, and a party of ten set out for the *Discovery*. The day was fine, and the travelling comparatively easy, though few of our crew were as yet so expert on ski that a trip-up and fall were not momentarily expected. However, the journey was pleasant and invigorating, and fair progress was made across the snow-covered ice.

When about one mile from Hut Point an excited party of perhaps twenty men came ski-ing rapidly out to us, and hailed us with the news that the southern party had just returned safely, having reached 82° 17′ south latitude. They had suffered much from scurvy, and all dogs had either perished or been killed.

This was the pioneer sledge journey to the south, and a remarkable feat under such adverse conditions, the three men being absent over ninety days from the ship. Lusty cheer-

THE VOYAGES OF THE 'MORNING'

ing was exchanged between the parties, and, hitching themselves to the traces, the *Discovery* men gave us a grateful help-in with the sledges.

Rounding Hut Point, here, at last, was the *Discovery*, completely frozen in. The little bay formed an ideal winter harbour, having, when ice-free, just swinging room for a ship at anchor.

The 'settlement' was quite animated and civilised looking. On Hut Point stood the roomy hut, with the magnetic huts close by. Adjacent to the ship were various rough shelters and dog kennels, and astern, the meteorological screens and anemometer. Sledges and ski stood near the gangway, or 'front door,' and sundry discarded cases, casks, and empty tins gave a homely aspect to the scene.

The southern party had received a tremendous ovation on arrival. They were sighted the night before, insignificant black specks on the Great Barrier, and a party went out to accompany them home. By the time we arrived at the *Discovery* they had indulged in a welcome bath, the luxury of a

CAPTAIN SCOTT LEAVING THE 'MORNING'

'DISCOVERY' IN WINTER QUARTERS

SCOTT AND HIS COMRADES

shave, and the comfort of a change from their dilapidated sledging gear.

In the ward-room we met Captain Scott, that gallant officer and ideal leader, to have known whom was a privilege. Even after all his recent hardships, sickness, and almost starvation, he was already interesting himself in the comfort and welfare of the others, and received us with affectionate hospitality. He was overjoyed to be 'home' once again, and delighted to know that the relief ship had arrived safely. After hastily glancing through his imposing looking mail, and learning also that the whole company had received good news, he said that it was a blessing to be able aptly to quote that 'joy cometh in the morning.'

Poor Wilson and Shackleton were confined to their bunks, so severely had the scurvy served them. The former's leg was badly swollen, and had not been improved by a fall in a crevass. Shackleton was very ill, and was, on the return journey, in a dying condition, but through dogged determination and indomitable spirit they had pulled through, and accomplished a journey which,

THE VOYAGES OF THE 'MORNING'

for sheer pluck and endurance, was difficult to surpass.

The whole world now knows of the subsequent tragic end of two of this party, who are justly ranked with the heroes of our nation. I shall never cease to be thankful for the great privilege of meeting and knowing such noble men.

But at the time of this narrative there was no lurking tragedy of sorrow and death, and that evening of February 3, 1903, was one of utmost thankfulness and extreme joy. Had not all sledge parties returned without disaster ? Was not a furthest south record established ? Were not all the results of the Expedition successful ? Were not hopes high that in the course of a few months all would be home again amongst their nearest and dearest ? And the relief ship had brought only good and cheerful tidings. Among such splendid men the feelings could not be happier.

In keeping with the prevailing joyous abandon, a ' banquet ' was held in the evening ; that is to say, an extra ' whack ' of New Zealand mutton was permitted, and

A FEAST AND A SPEECH

the fattest plum puddings were produced! The menu comprised canned soup and fish, mutton and potatoes, plum pudding, canned fruit, savouries, coffee, and a generous supply of wine and liqueurs, topped off with cigars. There were some who sipped their champagne from tin pannikins, and somewhat worn enamelled plates had to be occasionally introduced, but this jerseyed and mufflered party did not eat or drink the less because of these irregularities. The good cheer so predominated that even poor Shackleton made an effort to take his place at the table, although he could not sit a course through. Wilson did not attempt it; but the two did remarkable justice to the good things served in their cabins.

Stirring speeches were, of course, delivered, and Captain Scott, in reply, was almost overcome with the joy of it all. Think what it was to be suddenly surrounded with such luxury and relief after that awful ninety days' sledge journey! Even now, so impressed was I at the time, I can practically remember word for word his manly, modest response. 'You are all such jolly good fellows,' he said, 'that I can't thank you enough for the words you have spoken, and I can't hope to explain what it is to me to have men like you with me. But you can understand my feelings, I dare say, and I ask you to let me off speechifying, as there'll be no end of it to do when we get home, I expect. I should like to add that, under the circumstances, we cannot be too thankful that we have the relief ship with us.'

By this last allusion, it had evidently dawned on Captain Scott that, owing to the presence of so much ice so late in the season, it was possible that the *Discovery* would be detained for another winter.

It was astonishing the amount of food the three members of the southern party consumed! (Perhaps it was rude to notice this, but it was almost painfully apparent!) After a severe and lengthy sledge journey on reduced rations the longing for food is naturally great, and when it can be obtained in plenty the result is usually a gorge! Men get very sick after it, but go on eating and eating, until in a few days', time

HUNGER AFTER SLEDGING

the digestive organs return to their normal condition, and the ravenousness moderates. One was apt to entertain alarm at this awful gorging, and expect some terrible upset, but the doctors said that it was only natural after starvation, and no harm could result.

The evening was devoted to music and song. I could not help observing Captain Scott, who appeared restless in his happiness. One minute he would be comfortably lounging in an armchair smoking a large cigar and enjoying the music; the next he would suddenly disappear into his cabin to read some of his mail; then he would return again to the ward-room, remarking at each reappearance, ' By Jove, this is relief absolutely!' But he only too frequently visited the pantry to have another go at the remnants of the dinner! Then he'd surreptitiously steal into Wilson's and Shackleton's cabins with fresh supplies from the pantry! He was duly observed, but, knowing that this craving for food had to be satisfied somehow, no one pretended to notice the determined and continuous efforts he made to satisfy

THE VOYAGES OF THE 'MORNING'

the inner man, not only in his own case but also in that of his companions.

At midnight I went up to Hut Point with one of the officers, and, with two flags tied to broom handles, we managed to get the news through by semaphore to Evans, who was on the *Morning*. Although there was much mirage, he received the signal and replied, 'Heartiest congratulations from all on the *Morning*.'

We slept any and everywhere that night, the surplus hands in sleeping-bags on the cabin floors. I had not been asleep more than perhaps an hour, when I was awakened by hearing Captain Scott rousing Shackleton, whose cabin was next door.

'Shackles!' I heard him call. 'I say, Shackles, how would you fancy some sardines on toast?' In a little while the smell of toasting bread at the ward-room fire permeated the place, and a few minutes later I heard Wilson thanking the captain for the luxury! This continued at intervals during the early hours, and struck me as being at once humorous and pathetic.

CHAPTER X

Weather-bound — Sluggish ice field — Differences in seasons—Sledging over supplies—A coal depot—Sound freezing over—Farewell dinner on *Morning*—Modest requests—*Discovery* held for second winter—Departure of the *Morning*—A sad parting —' The Ice King '—Heavy new ice—*Morning* nipped —An anxious night—Our narrow escape—Sailing north — Days close in — Unpleasant weather — Arrival at Lyttelton.

WE were weather-bound for two days on the *Discovery*, which was a blessing in disguise from one point of view, in that it afforded an opportunity to learn, at first hand, the interesting details of the splendid work accomplished.

The ice field remained disappointingly sluggish during the following weeks, and only on one occasion was there any decided break-up, when for some unaccountable reason three miles of ice broke away in as many hours. The dense pack across the

111

head of the sound still remained, and could often be seen miraged above the northern horizon. Its presence there effectively prevented a swell from rolling in, which was now the only means of breaking up the five miles of solid ice between us and Hut Point. It was assumed that as long as the head of the sound was blocked the chances of its breaking away that season were small.

It will be remembered that in the preceding year, the *Discovery* encountered very little pack in this vicinity, which explains, no doubt, why the sound had been so free, and the ship had been able to penetrate, without difficulty, into her snug winter harbour. The theory put forward, that if the ship could get into the bay one year she ought to get out of it the next, was upset by this difference in ice-distribution. Therefore, although much optimism was evinced that the break-up would eventuate sooner or later, it was decided, after Captain Scott's return, to sledge over from the *Morning* the extra provisions, fuel, and clothing brought out for the *Discovery*.

Sledging, then, became the order of the

SLEDGING GEAR AND SUPPLIES TO 'DISCOVERY'

COAL DEPÔT ON GLACIER TONGUE

LANDING STORES

day, and continued strenuously through most of February. As coal supplies would be short after another winter, and as it was impossible, in the limited time, to sledge this commodity over as well, the *Morning* steamed alongside the handy glacier tongue off the Erebus slope, and deposited thirty tons, which stock could be drawn from, if necessary, the following spring. Evans, Doctor Davidson and I, and another party of three seamen, formed the transport, and sledged bags of coal all day to the upper ridge of the glacier. Dragging up the slope was stiff work, but the competition was keen, and we never permitted the others to outpace our record.

By the end of February no more ice had broken away, and as the open water had commenced freezing over at nights, it was expedient for the *Morning* to leave before she also got nipped. Our ship could do no more now. The *Discovery* had been re-provisioned and amply supplied with fuel. Under existing conditions only Nature could destroy the imprisoning ice, but as the head of the sound remained apparently blocked,

THE VOYAGES OF THE 'MORNING'

all hopes for the *Discovery's* release were practically dispelled.

Captain Scott held that blasting operations on an extensive scale might solve the difficulty, and in his dispatches he recommended strongly that a large supply of gun-cotton be sent south the following season in the *Morning*, provided, of course, that the *Discovery* was held for the winter.

The day fixed for the *Morning's* departure was March 2, which is as late as a vessel of limited power can remain in the sound. The season closes in very rapidly, and the sea now showed marked signs of freezing, with young ice forming regularly at nights.

The evening before sailing the entire *Discovery* staff, captain, officers, scientists, and crew, sledged over to bid us farewell. A large mail was brought, and of the staff of forty-six men ten were to return home. Shackleton, being considered not fit enough to remain, was to his great disappointment included among the passengers.

The combined crews made rather a crowd on the little *Morning*, but by dint of ingenuity in improvising extra accommodation, all sat

A CHOICE OF GIFTS

down to dinner together. After the repast music and singing passed away a pleasant evening. Outwardly all assumed happiness, but deep down was a feeling akin to melancholy in the knowledge that the parting wrench was so close at hand.

Towards the end of the evening chums forgathered in the various cabins, and the *Mornings* endeavoured to glean from the *Discoverys* what little gifts they would like to have brought back to them from the civilised world. They thanked us for the kindly thought, but, of course, no one required anything, and implied that they were amply satisfied with what they had. We would not be turned off like this, however, and after insisting, we eventually elicited the most modest and extraordinary requests. One man suggested a box of crystallised fruit. Another fancied a ' cadge ' pipe—' the largest one available, to use only when visiting a neighbouring cabin mate.' These and similar desires were subsequently gratified.

The following day was dull and overcast, with a biting wind and driving snow. This did not tend to enliven one's feelings. After

THE VOYAGES OF THE 'MORNING'

an early lunch the *Discovery* party got over on the ice with their sledges and dogs, and the ice-anchor was hauled on board. Then the *Morning* turned, and amidst stirring cheers steamed north. Cheer upon cheer went out, the whistle was blown, and the ensign dipped; but it was, nevertheless, a truly sorrowful parting. How small the little group looked, cheering at the ice edge! We were going back to friends and the comforts of civilisation in New Zealand. They were remaining to face the dreary Polar night, and for twelve long months they would be completely cut off from the outer world. But they cheered and cheered again, that plucky little band, until at length, as the ship proceeded northward, the cheering could no longer be heard. Then, harnessed to their sledges, after one pathetic final wave, they set off on their lonely way, and, marching southwards towards their winter home, were soon lost to view in the gloom and drift. It was a sorrowful parting, sad enough for those who went to the north, sadder still for those left in the lonely wastes.

Inspired by this parting, a song, 'The Ice

WORDS BY
J.D. MORRISON

The Ice King

MUSIC BY
G.S. DOORLY

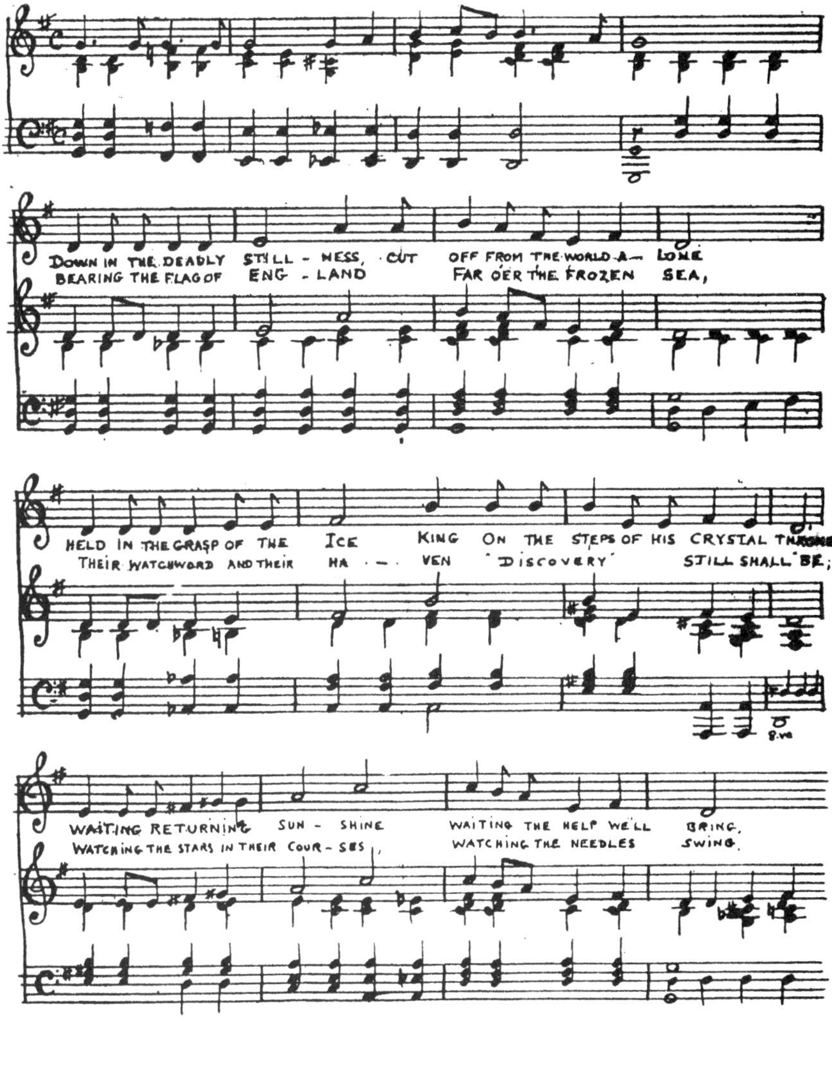

WEARILY WATCHING THE hours GO BY, TILL THE "MORNING" COMES WITH THE SPRING,
DOING THEIR DUTY NOT COUNTING THE COST, TILL THE "MORNING" COMES WITH THE SPRING:

REFRAIN:
PLAINTIVELY   *VIGOROUSLY*
FAR AWAY IN THAT COLD WHITE LAND IN THE HOME OF THE GREAT ICE KING, BRAVING HIS FURY DARING HIS WRATH WHEN HONOUR AND GLORY ARE

RELIGIOSO
SHOWING THE PATH; GOD WILL KEEP THEM FROM HARM AND SCATHE TILL THE "MORNING" COMES WITH THE

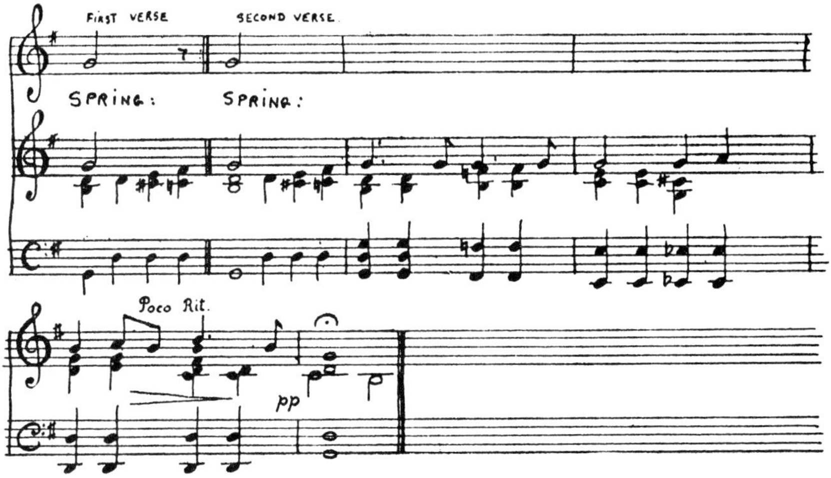
FIRST VERSE
SECOND VERSE
SPRING:
SPRING:
Poco Rit.
pp

THE ICE KING

King,' was written, and fittingly conveys the impression of that sad time :—

THE ICE KING

I

Down in the deadly stillness, cut off from the world,
  alone ;
Held in the grasp of the Ice King, on the steps
  of his crystal throne ;
Waiting returning sunshine, waiting the help we'll
  bring,
Wearily watching the hours go by, till the *Morning*
  comes with the spring.

II

Bearing the flag of England far o'er the frozen sea,
Their watchword and their haven, *Discovery* still
  shall be ;
Watching the stars in their courses, watching the
  needles swing,
Doing their duty, not counting the cost, till the
  Morning comes with the spring.

REFRAIN

Far away in that cold white land, in the home of
  the great Ice King,
Braving his fury, daring his wrath,
When honour and glory are showing the path ;
God will keep them from harm and scathe—
Till the *Morning* comes with the spring.

THE VOYAGES OF THE 'MORNING'

Here, then, was the *Morning* homeward bound, but, alas, without the cherished hope of her consort's company. The plans of the Expedition had now been materially altered. It was supposed that both ships would return to New Zealand, and, after the necessary repairs had been effected, that they would sail for England. Fate, however, decreed otherwise.

Owing to the *Discovery's* unavoidable detention, affairs assumed rather an embarrassing complexion, as the Expedition funds were not sufficient to meet the unexpected expense of at least another year. It was evident that under the unfortunate circumstances assistance must, somehow, be forthcoming to enable the *Morning* to return the following season, and if it were impossible even then to extricate the *Discovery*, her gallant crew, at any rate, had to be rescued. Nature was so subtle in her workings that even at this advanced stage in the season there still remained a chance of the ice breaking sufficiently to free the *Discovery*, but it was such a slender hope that little weight was attached to it.

The gloomy day wore on, and the *Morning*,

THE FROZEN SOUND

after steaming through extensive sheets of young ice, arrived at the head of the sound, to discover that more intense freezing had taken place there. The entrance, or, more correctly now, the exit, across from the Erebus shore to the western land presented an alarming looking sea of smooth new ice about six inches in thickness.

The *Morning* struggled bravely into this, but was soon brought up. With pack ice, comprised of brittle, broken-up floes and blocks, there is always a hope of chance openings to work a ship into, but with newly formed ice the smoothly frozen sea presents a sludgy and sticky substance which, by its adhesive nature, effectively arrests a ship's progress.

Efforts were made all night and the following day to push north, but with little result. With engines going full ahead for half an hour at times, the only appreciable effect was a raised mound under the bows of forced-up stuff, very similar to thick porridge. It was exasperating work, and our old bugbear, Beaufort Island, standing abruptly out of this smooth surface, two miles distant, seemed like a stolid sentinel, jealously guarding the sound. The ship was stopped for several hours and remained fast in the freezing sea.

Passing thoughts were neither cheerful nor encouraging. Had the ship been nipped there she would have suffered certain destruction, as our earlier experience of the dangerous currents in this locality only too clearly indicated. Granite Harbour was abeam, but so far distant as to be practically impossible to make under existing conditions.

But the gravest anxiety was the fact that nearly all fuel in the shape of oil, candles, and matches had, in sympathetic generosity, been given to the *Discovery*, so that the reflection of perhaps a forced winter in the Antarctic was inclined to be depressing. The position, certainly, did not have a bright outlook, and Nature alone had to be relied upon to solve the difficulty.

During the afternoon of the second day a blizzard could be seen approaching from the south, with drift driving before the wind in great clouds over the icy surface. The captain considered that, in our em-

NATURE'S SUBTLETY

barrassing situation, this opportunity should be embraced, and, risky though it seemed, ordered all sail to be set, and await results.

In a few moments the wind moaned fitfully, and rapidly increased into a strong steady breeze, with hard driving snow. With the first gusts the ship heeled slightly, and as the wind strengthened, the sails, bellying to their utmost, caused the braces and sheets to crack alarmingly with the strain, but the ship remained stationary. The engines were now making maximum revolutions, the masts bending like willows under the severe pressure of canvas, and as the blizzard grew fiercer and fiercer, with the ship straining terribly, headway was just perceptible. It was a desperate risk, but it was a tight corner. The ship forged ahead slowly, only too slowly, churning through the stodgy ice, and in the intensity of the anxious strain it was momentarily expected that the sails would blow out of the bolt-ropes, or, worse still, that the entire masts might go by the board.

Indifferent headway was made through one fearful hour; then Nature demonstrated

THE VOYAGES OF THE 'MORNING'

her subtlety in an eerie and unseen power, which caused the ship, regardless of the terrific forward strain, to go astern!

This has to be experienced to be realised. Imagine the force exerted by a driving gale against a full spread of canvas, accelerated by a forward thrust of the engines, to be thus calmly ignored by some fearful hidden agency! Man, undoubtedly, forms a part in some almighty scheme, but he cannot fail to realise how infinitesimal a unit he really is in this mysterious universe. Nature then permitted the breath to come freely once more by allowing the ship slowly to gather headway again.

A 'lane' of open water being discernible along the Beaufort Island shore, the ship was gradually and with much difficulty worked into it, and, repeating the Franklin Island evolution, by narrowly skirting the land, the north end of the island was safely passed, and, with utterances of heartfelt thankfulness from us, the new ice then encountered became appreciably thinner. For days a sea of filmy young ice of pancake formation was steamed through, unmistak-

MEETING WITH MR. WILSON

ably proving that the *Morning's* escape was only just accomplished in the nick of time.

The days were now closing in rapidly, and the presence of bergs in the darkness caused some anxiety; but, provided the weather was clear, these great floating dangers could readily be detected in the darkest night by the weird luminous effect they cast. Happily, the ship made good progress, and was soon clear of the Antarctic and its accompanying terrors.

Unpleasant weather conditions prevailed on the way north, but that was only the common lot of the seafarer; after experiencing the rigours of the Polar regions, gales in the more civilised latitudes appeared to have lost much of their dreaded sting.

In three weeks' time the *Morning* once again arrived at Lyttelton. A large passenger steamer was in company up the harbour, and, edging closely, a voice from her bridge hailed us saying that Mrs. Wilson was on board, and asking that her mail be detained. She had journeyed out to New Zealand to meet her husband, and it was rather odd

THE VOYAGES OF THE 'MORNING'

that her arrival should coincide with the *Morning's*.

As may be imagined, a great ovation greeted the little vessel, and the genuine expression of good cheer and welcome home from all quarters was in itself ample compensation for the trying experiences of the past months. Thus, the *Morning's* important mission of seeking and finding the *Discovery* was accomplished, and, although that ship herself had not returned, any feelings of doubt and anxiety regarding the welfare of her brave little band could now be dispelled.

CHAPTER XI

Winter in Lyttelton—Expedition taken over by Admiralty — *Terra Nova* purchased — Incidents, humorous and otherwise—H.M.S. *Phœbe* assists in *Morning's* overhaul—Second departure from New Zealand—Tasmania—*Terra Nova* at Hobart —A unique tow—Transhipping stores—Sixty-third anniversary of *Erebus* and *Terror* at Hobart— Hospitality—An original dancer—Scuffle with a sailor—H.M.S. *Royal Arthur* supplies gun-cotton —Departure from Hobart.

THE *Morning* spent the winter months at Lyttelton. Our cabled news caused mixed feelings: delight that the *Discovery* and crew were all well and had performed such good work, but disappointment at her detention and the consequent urgency for funds. The cabled reports from England were so conflicting and unsatisfactory that, soon after our arrival, the captain proceeded home to explain the situation and advise on the best means to meet the difficulty.

THE VOYAGES OF THE 'MORNING'

The British Government ultimately took over the Expedition, and, under the direction of an Antarctic Committee formed at the Admiralty, purchased the Newfoundland sealing vessel *Terra Nova* to act as an extra relief ship. On the theory that two ships are safer than one, this ship was sent to accompany the *Morning*, and Captain Colbeck of the *Morning* was placed in command of the Relief Expedition until its arrival at McMurdo Sound, when all ships would be under Captain Scott's jurisdiction. This is the same *Terra Nova* that was employed on Captain Scott's last expedition seven years later.

Our officers stayed for some time at a hotel in Christchurch, where we received the greatest hospitality from the proprietor and his family. One section of the hotel, allotted to us, was known as the *Morning* corridor. An institution was formed, for tea and toast to be brought up to Shackleton's room every morning, and each man met there for this eye-opener. Apropos of this, we had been to a dance one evening, and one of the officers had intended returning to

SHACKLETON ON ARRIVAL AT LYTTELTON

A STRANGE TOILETTE

the ship to sleep, because his various articles of clothing, used at the hotel, were at the laundry. The last train had gone, however, before the dance was over, and he was forced to remain in town. Imagine the effect produced when this fellow sleepily arrived in Shackleton's room next morning for his usual tea and toast clothed in a lace-trimmed nightdress! Tea-cups, &c., flew in all directions, and an undignified rush ensued round the corridors after the unfortunate chap. Such a sensational exhibition naturally astonished, and yet apparently amused, the housemaids. Before turning in, this officer had looked through the wardrobe and chest-of-drawers on the off chance of a sleeping-suit being overlooked, and, strangely enough, this garment was discovered. It had been left in a drawer by one of the maids.

Shackleton was practically well on our arrival in New Zealand. The night before he left for England a farewell dinner was given, and is a pleasant recollection.

The ship was laid up for the time being, and a general overhaul commenced. At all times during the day interested little groups lined the wharf to view the ship that had actually been a voyage to the Polar regions, and about which so much had been said in the local papers. Remarks of a more or less flattering nature were often overheard, but on one occasion at any rate there was a diversion. A dilapidated old man was standing on the wharf and sucking at a very worn stumpy clay pipe. After listening to the conversation of the crowd for some time, he at length broke his silence by remarking : ' 'Eroes you call 'em ? ' he said. ' 'Eroes ? ' (I regret to record that at this point he contemptuously expectorated on the wharf.) ' 'Eroes ? Why, the last time the beggars was 'ere they got my girl that dilly that she give up the bloke she was keeping company with ! '

Evans was full of original ideas, and the following incident is characteristic. Some ladies on board to tea one afternoon noticed with interest the ' barrel ' at the mast-head. One girl said that she wouldn't mind going up to it, whereupon Evans took her at her word and offered to find another girl who could go up to the crow's-nest and down in quicker time. The contest was accordingly arranged, and on a bright moonlight night soon after-

A LADIES' RACE

wards this feat was actually performed. A small party comprising the competitors, with their respective chaperons, assembled. The girls, attired in regulation gymnastic costume, left the deck at a given signal, and, each on her own side, climbed the rigging with wonderful activity. They reached the crow's-nest, got inside, and down again on the opposite side of the mast. It was done in a very short time, and there was little to choose between them. An impromptu supper brought a rather original evening to a happy conclusion.

As we were to spend the winter at Lyttelton, Evans applied to join one of the vessels on the station to keep up to date in his naval work. He had recently been promoted to lieutenant, and was appointed to H.M.S. *Phœbe*, then in New Zealand waters.

The Admiralty cabled to the commanding officer of the station to despatch a man-of-war to Lyttelton to assist the *Morning* in preparing for her second southern voyage, and fortunately the *Phœbe* was detailed for this. Evans, who had only been a few weeks away, was in the unique position of representing the two ships. He boarded the *Morning* in charge of a party of bluejackets, to overhaul rigging, etc., and eventually found himself tallying stores against me in the hold! The work was thoroughly done, and I, being in charge of the victualling department, particularly appreciated this help.

We attended innumerable social functions, and were always most hospitably treated, and even free passes on the railways were furnished by the New Zealand Government.

An incident I often think of occurred one evening when Evans and I were chatting in the hotel billiard-room. He had been returning calls that afternoon, and amused me very much by describing his experiences with a pony and trap he had hired for the occasion. Like most sailors, he knew little of horsemanship, and his efforts to go 'hard-a-port' or 'hard-a-starboard' did not come off as readily as he had expected. Leaving one house, he had to drive down a long carriage sweep bounded by grassy banks. His friends stood in the porch to see him away, and, much to his consternation and their amusement, the pony bolted up one of the banks, and, as he declared, ' deliberately tried to climb a tree!'

A 'STRONG MAN' FLOORED

This being a race meeting week in Christchurch, there were many country visitors in the billiard-room. Our conversation was rudely interrupted by a burly man who, coming up behind Evans, put his arms round him and squeezed him with unwelcome familiarity. I believe the man knew who we were, but I had not seen him before. Evans, of course, couldn't see him at all then, but strongly resenting this, he suddenly wheeled about, and in a moment the man measured his length on the floor. It was cleverly done, and caused a scene. Evans fortunately controlled his temper sufficiently to avoid striking a blow; in fact, he calmly grabbed the man by the waistcoat and lifted him to his feet again. He then quietly told him not to be so familiar in future. Totally crestfallen, the man slunk out of the room. The company then cheered. They explained that the man fancied himself as a strong man and amateur pugilist, and always 'talked big' of his prowess in that direction. They were therefore pleased to see his conceit subdued, and especially by one not much more than half his size.

Amongst our varied social amusements

THE VOYAGES OF THE 'MORNING'

was hockey, which we often played during the winter against a ladies' hockey team. One of the club's members, Miss Russell, who was also the team's goal-keeper, lived close to the playing field. Her people very kindly extended a standing invitation to Evans and me to lunch with them on match days. As time went on they became so hospitable and generous that we regarded this place as 'home' and popped in at any time. This home life was naturally much appreciated.

Having some relations in Dunedin, I paid them several visits during our stay in New Zealand. This brings me to record another coincidence in my associations with Evans. Oddly enough, within one week of each other, he and I became engaged to be married, he to Miss Russell in Christchurch, and I to a friend of my relations in Dunedin.

In due course all arrangements for the Expedition were satisfactorily concluded, the only disappointment to us being the Admiralty's decision to send the *Morning* to Hobart, Tasmania, to sail south with the *Terra Nova* from that port.

Towards the end of October 1903 the

AN UNLUCKY VOYAGER

ship was once more ready for sea. About a week before leaving our chief officer, England, met with an accident by a dray-wheel passing over his foot, which very nearly caused him to be left behind. Through some extraordinary means England was frequently 'in the wars,' and the wonder of it was that he was never killed. We were not a hard-hearted lot, but somehow his misfortunes always seemed to have a humorous side. One day, with about a minute to catch a train, he rushed over the ship's rail for the shore. Unfortunately he missed his footing and fell into the water between the ship and the wharf. He had his evening dress in a bag with him at the time! He put three hired bicycles out of action during our stay in Christchurch, through collision with some vehicle or lamp-post! But his most astonishing accident occurred at the hotel. The piping under the roof-coping became blocked with leaves, and, being a most obliging man, he promptly volunteered to clear it. He shinned up the drain-pipe, and, when only a foot or two from the roof, the slender piping, not intended for such usage, naturally gave way. Down came England with a terrible crash on to a stained-glass skylight over the main passage of the hotel! With a leg sticking through each side of the framework, and the hall-way littered with broken glass, it was indeed a pathetic picture. He had on a new suit of clothes which was torn in places, but he himself wonderfully escaped injury.

Leaving Lyttelton was now like leaving home. We had made life-long friends, and the parting was hard. Another thrilling send-off was accorded us as we steamed down the harbour. Several long months must now be passed before we should again be with our good friends, and as the lot of those who remain is the harder, they must have put in many anxious hours of waiting and watching.

The passage from Lyttelton to Hobart was moderately fine and devoid of unusual incident. Early in November we steamed up the Derwent River and were charmed with the beauty of the approaches to Hobart, with Mount Wellington towering above the picturesque town. On arrival, we found

'ENGLAND'S DOWNFALL'

AFTER SIXTY YEARS

that the *Terra Nova* had preceded us by a few days. Being unavoidably late in leaving England, the lost time was made up by men-of-war towing her. This was performed by three men-of-war respectively, from Portland to Gibraltar, Gibraltar to Suez, and from Suez to Socotra. Experiencing fine weather, she then proceeded, unaided, across the Indian Ocean, and reached Hobart in record time. As far as I remember, this unique voyage occupied less time, actually, than the regular mail service from England to Australia !

The *Terra Nova* brought a large shipment of stores for both *Discovery* and *Morning*, and a considerable time was spent in sorting out and transhipping.

History was repeating itself. Sixty-three years before, to the month, two Antarctic vessels were at Hobart. These were the famous *Erebus* and *Terror*, under the command of Captain Sir James Clark Ross, R.N. He sailed on November 12, 1840, for the South Polar regions, and discovered that portion of the Antarctic continent which was the field of the present Expedition.

THE VOYAGES OF THE 'MORNING'

We were very interested to meet an old identity in the Government service who remembered the *Erebus* and *Terror* at Hobart.

To commemorate the anniversary of Captain Ross's departure from Hobart, the Mayor and several leading citizens entertained the officers of the *Morning* and *Terra Nova* to a cruise and picnic up the beautiful Derwent River.

It may be interesting to mention here that, on our return to England the following year, we had the honour of meeting the sole survivor of the *Erebus* and *Terror* Expedition. This was the late Sir Joseph Hooker, F.R.S., the eminent botanist. In Ross's Expedition there was no such thing as photography. As an example, however, of how the mind is impressed by the awe-inspiring fascination of the Antarctic, any photograph shown of mountain, cape, or island was readily named by Sir Joseph.

The people of Hobart were extremely good to us, and spared no pains to make our visit a welcome one. On the King's birthday we attended a levee at Government

AN ORIGINAL DANCER

House, and the Governor, Sir Arthur Havelock, presided at an official reception tendered to us. Although not exactly the season, several dances were arranged in our honour.

Owing to his accident in Christchurch, England reluctantly missed several of these functions. A dance given by some special friends of his was, however, too much for him to resist. Strangely enough, he was the only representative from our ship. Some of us were working on board that night. The idea of England's going to a dance with a game leg and a crutch was too much for us, and our curiosity was so much aroused that we went along to see how on earth he was getting on. Stealthily groping our way through the garden, we could see the revelries through a large open door, without being seen ourselves. The Lancers was in progress, and to our intense amusement we caught sight of England, 'going strong,' hopping round and round on one leg, with ⁓tick in his hand! Our laughter we were discovered, consternation, we appeared on the scene. For many months afterwards his leg, although better, suffered a good deal of pulling.

The *Terra Nova's* crew was comprised chiefly of Arctic whalers, rough and hardy men. For some reason they were not very friendly with our crew, though the antipathy wore off later on in the common hardships of the South.

I have to record yet another pugilistic encounter in my associations with Evans. We were working late on board one night, and had just turned in. The other officers were on shore. In a little while we heard a man on the wharf using some very bad language, and the gist of his foul remarks was that the *Morning* and her crew were not worth a ——! We put up with it for a few minutes, when Evans suggested that he ought to be suppressed. We felt certain that he was one of the *Terra Nova's* men, and rightly guessed him to be a very powerful man known as 'The Life-Guardsman.' He had been a soldier once. We went on deck as we were, in pyjamas, and saw the man standing on the wharf alone, close to th

A SCUFFLE WITH A SAILOR

ship's rail. We told him to go back to his own ship, whereupon he swore dreadfully and defied anyone to move him. As we approached he drew out a knife attached to a lanyard round his neck. This was too much to overlook, and we sprang at him together. The man had, of course, been drinking. In spite of his strength, he couldn't hope to get away with two sober and energetic athletes. He was soon flat on his back, and got a severe thrashing. Our watchman, alarmed by the scuffling, jumped on to the wharf with a lantern and illuminated the scene. When the worst struggles had subsided Evans took the man by the shoulders and I by the legs, and we dragged him along the wharf to the *Terra Nova.* Swinging him then, in a one, two, three fashion, we hurled him over the rail, and he dropped with a thud on the deck of his ship. The unavoidable disturbance awakened the *Terra Nova's* captain, who also appeared in pyjamas. We apologised for dealing so severely with one of his crew, but the captain, being used to whalers, promptly ordered him to be locked up for the night. I don't think ' The

THE VOYAGES OF THE 'MORNING'

Life-Guardsman' ever knew who, or what, struck him!

H.M.S. *Royal Arthur*, then flagship on the Australian Station, arrived at Hobart just before our departure, and supplied both ships with a large quantity of gun-cotton. This was in accordance with Captain Scott's request, to facilitate the freeing of the *Discovery*. By the first week in December the ships were fully loaded and ready for sea.

After a 'send-off,' almost as enthusiastic as from New Zealand, the ships left Hobart on December 5, 1903. We anchored for the night, and sailed finally at dawn next day. As the ships steamed out, the breaking day over the peaceful harbour and surrounding beautiful hillsides made one of the finest pictures I have ever seen.

CHAPTER XII

Southward—*Terra Nova* and *Morning* in company—A narrow shave—The ice pack again—Second Antarctic Christmas—Scott Island sighted—'Dead reckoning'—Admiralty Range—Deceptive distances—Fascination of the Antarctic—A quaint group—Sea-leopard hunt—View of South Victoria Land—Franklin Island—Difference in pack distribution—McMurdo Sound—Sighting *Discovery*—Extensive sheet of field ice—Hopeless prospects.

VOYAGES to the Antarctic through the stormy Southern Ocean can never be called pleasant, but the second voyage to the Far South was at any rate less lonely, owing to the comfort of the *Terra Nova's* company. We steered away to the S.E. from Tasmania, and kept within a few miles of each other throughout the voyage, exchanging noon positions daily.

The *Terra Nova* was a better steamer, and could always overhaul us, but under sail alone the *Morning* had the advantage. Only once during the thirty days' journey to McMurdo

THE VOYAGES OF THE 'MORNING'

Sound did the ships part company, and that was for a few hours one night during a heavy gale.

The *Terra Nova*, much to our secret displeasure, used to describe a circle round the *Morning* each twenty-four hours, clearly demonstrating her superior speed. At dawn she was on our quarter, by noon abreast of us to signal her position, and at dusk ahead somewhere, on one or other bow.

One afternoon, thinking, I suppose, that they could always 'make rings' round the little *Morning*, the *Terra Nova's* people endeavoured to run their ship close across our bow. This nearly resulted in a serious mishap. A fresh breeze was blowing, and both ships were under sail alone. The *Terra Nova* being the weather ship was the giving-way vessel, but she suddenly headed across our bows. It is imprudent to do this at any time, and especially in this circumstance, when both ships were under sail and the *Morning* was known to be the faster sailer. She bore down on us so closely that we had to make a decisive alteration in our course to avoid a collision. The more we kept away the more, for some

A CLOSE SHAVE

extraordinary reason, did she head us off, until we had reached that point when we were almost 'caught by the lee.' She skimmed across our bows by the barest margin, or, more correctly, by no margin at all, because our bowsprit grazed her port quarter-boat. A very little closer and our bowsprit might have been carried away, which would have caused more disastrous results. Carcasses of mutton hung on the span between the *Terra Nova's* boats' davits, and the humorous chief officer of that vessel, standing on the poop, seemingly unconcerned about anything else, shouted to us to 'mind the mutton.' Fortunately there was no accident, but it was a close shave.

On December 23 the first ice was sighted, and on Christmas Day the pack was entered. The pack was, therefore, met considerably further north than experienced the previous year, and this accordingly set us speculating. Either the season was earlier, and consequently more advanced, or the winter had been more severe, resulting in a wider pack area.

Christmas Day was spent in the customary fashion, and our good friends in New Zealand had supplied the ship's company with many

THE VOYAGES OF THE 'MORNING'

welcome delicacies. Before leaving New Zealand Evans's fiancée gave me a number of letters to hand over to him on certain dates, and the first was a Christmas greeting. At intervals throughout the whole voyage I had letters for him, and always placed them for him to find unexpectedly. My little joke was never to hand one to him personally, and, although he was pretty certain that I was in charge of them, he could never really tell, because I always assumed ignorance. On one occasion I was rather exercised in my mind how to get a letter delivered to him on a certain date. He was away sledging, and at that time the distance was so great to the *Discovery* that a journey to a half-way camp and back occupied two days. I managed my little duty by giving the letter to a sailor in his sledge party, and the following morning during breakfast at the camp he gave it to Evans. It was a quaint idea, but it had given Evans's fiancée pleasure to work up the little scheme, and I was glad to be able to carry it out satisfactorily.

On Christmas night Captain McKay of the *Terra Nova* and his chief officer dined on the

TO BED BY DAYLIGHT

Morning. The pack was not very dense, but the fog had set in, and we were waiting for it to clear. Captain McKay entertained us very much with tales of the Arctic and exciting whaling experiences. It was the wee sma' hours before the meeting broke up, but as it was now perpetual day one felt reluctant to go to bed in the daylight. Perhaps this may be more my own feelings. I was born and lived some years in the West Indies, where twilight is unknown. The day is practically from 6 A.M. to 6 P.M. My parents first took me to England when I was eight years old, and on arrival we stayed for a few days at Plymouth. It was about the middle of June, and I can remember how puzzled I was to know why it was necessary for me to go to bed before the sun had set. My elder brother and I used to creep out of our beds and gaze longingly and sadly at the people walking about on the Hoe while the band played, quite ignorant of the fact that it was, perhaps, nine o'clock at night. My first impression of England has been a lasting one, and, despite the childish disappointment of spending daylight hours in bed, I have always loved going to Plymouth.

THE VOYAGES OF THE 'MORNING'

After working two days in the pack Scott Island was sighted, about eighteen miles to the eastward. We were quite satisfied to view it from a good offing, as we had not forgotten our narrow escape from disaster there a year before. In the previous season the pack was met further south even than Scott Island. This year dense pack extended well north of it. Although heavy pack was occasionally forced through, it was apparent that the present season's ice was not as dense as the last. Much more fog and falling snow were experienced in the pack, and for five days the sun was obscured. The ship's position during this time was determined by 'dead reckoning,' as it is called, and a difficult thing to gauge owing to the erratic courses and distances made. The sun at last broke through on December 30 and cleared away all fog. By observation we found, to our satisfaction, that the ship's position was only fifteen miles out from our reckoning, not at all bad work for five days' approximation.

On New Year's Day (1904) the pack was cleared, and that evening the Admiralty

THE CALL OF THE SOUTH

Range was observed, distant about one hundred and twenty miles. Distances are deceptive in the Antarctic. In clear weather the rarefied atmosphere, assisted by the earth's flattened surface and the great altitudes of many of the mountain ranges, enables one to see peaks at astonishing distances.

It was fascinating to see these wonderful mountains again. We had only seen them on our search cruise along the coast the year before, and yet there was no difficulty in picking out and naming each peak. There is undoubtedly a wonderful attraction in these regions, whatever it may be. Men go to the remote corners of the earth, suffer untold privations, and face terrible risks; yet, knowing what is awaiting them, they return repeatedly. It may be to satisfy the desire to overcome obstacles which asserts itself, or a fascination may lie in endeavouring to unravel hidden mysteries. I am satisfied in my own mind, after two cruises in the Antarctic, that a charm certainly exists, which, perhaps, can only be described as 'the call of the South.' Alpinists have this fascination. The call to the mountains is as irresistible to them.

So little drift ice was passed after clearing the pack that it was quite easy to proceed for long stretches under sail alone. Of course we were well off shore and making direct for McMurdo Sound, assuming that the *Discovery* must still be there.

One day we came upon a large, rather rotten floe on which was a unique Antarctic group. It comprised three sea-leopards, ten crab-eaters, four or five giant petrels, numerous penguins, skua gulls, snow and brown-backed petrels. Possibly the scarcity of ice caused so many animals to gather on the one floe. The ship ranged alongside the floe, and naturally a general scattering ensued. The sea-leopards, being very fine specimens, were the chief object of our attention. Several of us blazed away with rifles, and, strangely enough, only one died on the floe. This was Evans's mark, and he and Morrison went away in the boat to secure the prize. The floe, being in a rotting condition, was difficult to get on to, and a swell lapping against it did not improve

PURSUED BY A WHALE

matters. In due course a rope was slung round the animal, and being of such an enormous size and weight, it had to be towed to the ship. The boat was then ridiculously 'by the stern.' Evans and Morrison struggled with the oars, and the boat's bow, being in the air, presented a most comical picture. But the most exciting part of the leopard hunt was the sudden appearance of a huge rorqual whale, which followed the boat, evidently interested in the dark object being towed in the water! I don't know whether rorquals attack seals or not, but the presence of a great whale under the stern of the boat was not at all appreciated by the oarsmen, who were expecting their end momentarily!

Clouds hung low over Coulman Island as we passed it, but the lower portion of the island was clear enough to show that there was no pack ice banked up against it. This was ample proof of the difference in the ice distribution this season.

The next day was beautifully clear, and the mountainous aspect of South Victoria Land was striking. To the south of Wood Bay the extinct volcano, Mount Melbourne, stood out clearly, and the land was visible to the northward nearly as far as Cape Adare.

Early on January 4 Franklin Island was sighted, and by noon the ships stopped half a mile off the Penguin Beach. The contrast between the two seasons was very marked. Strenuous efforts with heavy pack, and a providential escape from the island, attended the last visit, but on this occasion no ice was in sight. The day was perfect, not a cloud in the sky, and the temperature rising to 40° was comparatively warm. A party landed and collected penguins' eggs and rock specimens. So clear was it that Mounts Erebus and Terror and Beaufort Island were visible to the southward.

As conditions appeared so favourable, we shaped a course from Franklin Island direct to McMurdo Sound. It was intended to call at Cape Crozier first for news, but there was no reason now why we should. We could return there if by any extraordinary chance the *Discovery* had left the winter harbour.

On approaching Beaufort Island loose pack was met. The floes were very heavy and

CHANGED ICE CONDITIONS

hummocked, but the ships worked through and into the Sound without much difficulty.

On January 5, therefore, McMurdo Sound was entered, but progress was retarded by occasional belts of heavy pack. Soon the old familiar landmarks, Castle Rock and Observation Hill, were seen. The first sign of civilisation was a flag on a staff on one of the islets south of Cape Royds.

Soon after noon the *Discovery's* masts were seen over Hut Point. It was hard to imagine that nearly twelve months had elapsed since we were last there.

But there was one rather upsetting condition since our last visit. The *Discovery* was then imprisoned by field ice extending about five miles to the north. To our dismay we were now brought up against the field at a distance of about eighteen miles from Hut Point. The field was not rotten-looking either, and was eighteen inches thick at the outer edge. What, then, was its thickness further in ? Our disappointment was intense, and hopes fell to zero. The good old *Discovery* was evidently to be a fixture in the Antarctic for all time. Our arrival in the sound three weeks earlier in the season was the only slight consolation, but the prospects were apparently hopeless all the same. With much eagerness and anxiety, therefore, we awaited the first arrivals from the *Discovery*.

CHAPTER XIII

Captain Scott and Dr. Wilson—Astonishment at two ships—Doubtful chances of freeing *Discovery*—Evans and I at Scott's camp—An appetising ' hoosh '—Charm of an apple—Dangerous moving pack—Preparing to abandon *Discovery*—Sledging valuables to relief ships—Anxious days—Blasting operations—Snow blindness—Sailing orders issued—Depressing thoughts—Gale in Ross Sea—Extensive break-up of field—Visit to *Discovery*—Influence of swell at *Discovery*—Strenuous blasting—Sudden break-up of ice—Exciting rescues—Stirring hours—A struggle for priority—Dramatic arrival in winter harbour.

BEING eighteen miles distant from Hut Point, it was difficult at first to know just what to do. Sledges were prepared and mails and parcels sorted out ready for a journey across to the *Discovery*. A sharp look-out was kept continuously on Hut Point for an approaching party, but there were no signs of movement in that direction.

About 4 P.M. two men were seen approaching the *Morning*. They might have been from

THE VOYAGES OF THE 'MORNING'

the *Terra Nova*, but on close scrutiny through our best telescope we could tell by their general attire that they were *Discovery* men. But how they had travelled eighteen miles in this short time was puzzling to us. In a little while they drew near to the ship. I lowered a ladder from the bow to the ice, and they scrambled on board. It was not until they reached the rail that I could properly recognise our old and admired friends Captain Scott and Dr. Wilson. The joy at seeing them again was great, and for the moment, through some emotion or other, words of greeting refused to come. But Captain Scott, with a telescope under his arm, was ever the typical naval officer, in spite of his ragged and weather-beaten appearance. Shaking hands warmly, he asked : ' Is Captain Colbeck on board ? ' He might really have been calling on the captain of another ship in a squadron. Dr. Wilson was more weather-beaten. As soon as they got down on the deck the whole crew mustered and gave three ringing cheers. The visitors then shook hands all round in thanks for the stirring welcome. Captain Scott went aft to the chart-house

CAPTAIN SCOTT AND CAPTAIN COLBECK

A VISIT FROM SCOTT AND WILSON

with our captain, and Wilson had a very interested audience in the ward-room.

They were camping only four and a half miles away, at Cape Royds, watching the ice edge to see what breaking-away was taking place. Turning out that morning, they sighted the two ships half-way down the sound, and at first they could not believe their eyes. Why two ships, they wondered ? They expected the *Morning* to return some time that month, but what was the other ship for ? At first they thought it must be an optical illusion ; then perhaps it might be a stray whaler ; or, more likely still, the German Expedition ship *Gauss*, which was in the Antarctic then, and possibly had met the *Morning* and accompanied her. But the correct meaning of the *Terra Nova's* presence never for a moment entered their minds.

Ever thoughtful of others, they immediately set off for the ' half-way ' camp, ten miles south, close to the Delbridge Islets, and instructed the party there to take their sledges to the *Morning*, so as to get the mails away to the *Discovery* that day. Tramping back another ten miles, they

THE VOYAGES OF THE 'MORNING'

arrived at the *Morning*, as mentioned, about 4 P.M.

Being in charge of the mails, I was soon diving into the bags for Captain Scott's and Dr. Wilson's letters. The mail-room was just off the chart-house, and I could not help overhearing the conversation between the two captains. I was delighted to hear that all were well on the *Discovery*, and that some remarkable sledge journeys had been made. Our captain asked what the prospects were of the *Discovery's* being freed that year. Captain Scott replied—to use his very words—'Not a cat's chance.' This was very disquieting news to hear; but, as he explained, he had been camping on the shore close by for the last four days, and not a yard of ice had broken away in that time. He estimated the field to extend for eighteen miles from where we were to Hut Point. The *Discovery's* chances of getting out looked very remote, far more even than last season, when only five or six miles detained her. Our report of the open water in the Ross Sea, however, cast a small ray of hope on the otherwise gloomy outlook.

THE 'DISCOVERY' TO BE ABANDONED

When Captain Scott received the despatches from the Admiralty he was naturally much upset. The instructions were that if it were impossible to free the *Discovery* she must be abandoned, and her crew, with all records, instruments, specimens, etc., were to be brought back by the two relief ships. The *Discovery* had been a home to them for over three years, and it need hardly be said that not one man was willing to abandon her. Matters had looked so hopeless before our arrival that parties had been, and were still, out securing seals and penguins for another winter's food. But these orders altered their plans.

The *Discovery* was, I learned, encased in ice fourteen feet thick, as compared with ten feet when we left her last year. All sledge parties were back. For a fortnight the entire ship's company, including scientists, were at the 'half-way' camp, engaged in sawing a way through the ice. This work was kept going continuously in watches night and day, but as the results were so trifling sawing operations were abandoned. The ice cut through was seven feet thick.

THE VOYAGES OF THE 'MORNING'

As Captain Scott and Dr. Wilson stayed on board that night, Evans and I were permitted to go over to their tent for a day or two. We packed some provisions, including a few excellent Tasmanian apples. Captain Scott was amused at this, and told us that we could give a great treat to one at least of his men, who was due to arrive with provisions next day at his camp. On his recent sledge journey towards the Magnetic Pole over the western mountains Captain Scott was accompanied by Petty Officer Evans (now, alas, also dead) and Stoker Lashly. On their return they were weatherbound for several days in their tent by a severe blizzard up the mountains, and, as food was scarce, the outlook was bad. Captain Scott asked his companions one day what they would fancy most at that moment if their wish could be gratified, and Lashly promptly replied : ' A good old English apple, sir ! ' P.O. Evans's request was also for some trivial thing, and Captain Scott fancied a sixpenny jar of Devonshire cream ! The simple tastes of men in such sore straits is marvellous. Therefore, when Captain Scott

CAPTAIN SCOTT'S TENT, CAPE ROYDS

ARRIVAL OF MAIL FROM 'DISCOVERY'

[See p. 165

THE TALE OF AN APPLE

saw the apples he told us this story, which we bore in mind.

Evans and I set out about 1 A.M. and reached the camp at Cape Royds about 4 A.M. Hundreds of penguins surrounded the little green tent and kept up an incessant cawing. This was disconcerting until one became used to it. We had a warm drink and turned in. After two sledging seasons the sleeping-bags were not exactly beds of roses, and the reindeer hair which lined them inside was falling out. Every article in the tent, indeed, seemed associated with reindeer hair, and a very annoying discomfort was to awaken occasionally with a mouthful of the nasty stuff!

Well on in the forenoon, while we were still in our bags, we heard voices a long way off. Surmising this to be the party with stores for the camp, we lay low, like Brer Rabbit, and awaited their arrival. Crunch, crunch came the footsteps along the snowy surface, closer and closer, and, on reaching the top of the slope, the party hailed. As we made no reply, they opened the flap of the tent, and were greatly astonished to see the two new occupants. P.O. Evans and Stoker Lashly were two of this party of four. We turned out, rolled up the bags, and the six of us managed to crowd into the little tent. We sat round the primus stove, which Lashly soon had under way. A few strips of bacon, mixed up with stale oatmeal and seasoned with a generous amount of onion powder, soon sizzled in the cooker. The aroma was particularly pleasing, and we all dug into the 'hoosh' with much relish, despite the unavoidable fourth ingredient of reindeer hair! The snow in the outer cooker soon thawed and boiled, and tea followed. When the 'wreck' was cleared away, and everything put ship-shape, Evans groped into his sleeping-bag and produced a large apple. Words absolutely failed Lashly! It was quite pathetic. His eyes grew larger and larger as he gazed from the apple to Evans, and from Evans to the apple. I expect he thought it was an illusion of some sort. Fortunately we had an apple for each man, so there was no disappointment. Lashly fondled his apple lovingly for a long time before eating it,

PERVERSE CURRENTS

but when he did eat it, he ate it thoroughly, nothing being left—skin, core, pips, all went. Lashly considered this quite a red-letter day in his life.

The party now left for the *Morning*, and that evening Dr. Wilson returned, and made us an excellent meal of fried seal liver. Evans and I rambled all over the cape during our stay, and collected granite stones and penguins' eggs. The ' 'Evanly Twins' returned to the *Morning* next day, having had a very enjoyable spell at Cape Royds. We took two baby penguins back for our doctor.

During the first few days the field ice remained stationary. The only ice movement was the heavy pack, through which we had passed at the head of the sound. A current setting south during this time was indicated by the movement of the heavier floes and small bergs. Despite the prevailing south-east winds, they frequently set into the sound at a considerable rate. It was uncanny to see these great floating masses ' working to windward ' without any difficulty. As a result of this both *Terra Nova* and

THE VOYAGES OF THE 'MORNING'

Morning received some nasty squeezes against the field edge. Occasionally we had to get under weigh to avoid the heavier pieces. One night a continuous pack set into the sound. After rounding Cape Bird, it set directly to the ice edge, then along the edge for some miles, and eventually disappeared, setting to the north-west on the other side of the sound. There was no protection for the ships, and we spent a most unpleasant time with this dangerous stream of ice rubbing along the ship's side. Being against a solid body on the other side, there was no give at all. The ships, therefore, had to stand it as best they could, and at times the severe pressure caused the stout oak beams to buckle. We were much relieved when this uncanny motion ceased. By measuring the thickness of the field ice at intervals of a few days, it was proved that it was appreciably thawing. Cracks appeared from time to time, and large strips of ice broke away occasionally. This meant much trying work for the ships in continually steaming round and re-anchoring to the new edge. Very severe blizzards made matters worse on the

BLASTING OPERATIONS

ships, but effectively kept the sound free of dangerous pack.

As the ice was going out so slowly Captain Scott soon decided to sledge all specimens, instruments, etc., to the relief ships. The *Discovery* men sledged the stuff from their ship to the half-way camp, and the loads were brought the rest of the way by the relief ships' parties. The distance during the earlier stages necessitated a two days' journey. Evans was in charge of the sledge parties from our ships, and I had the less interesting work of loading, and discharging the sledges on their return.

After the first week blasting was commenced. The explosions blew out large areas with apparently little result. From the great concussion, however, cracks generally appeared, and an ocean swell was required to completely sever the cracked portions. In the previous season the head of the sound was so densely blocked with ice that no swell was felt during our stay. This year, however, so little pack was met with outside that our hopes ran high that a swell would occasionally find its way into the sound.

THE VOYAGES OF THE 'MORNING'

During the whole of January the ice was cracked by blasting, and slight swells coming in at times gradually broke away ten miles of the field. By January 31 we were, therefore, eight miles from the *Discovery*, and on that day we received a note from Captain Scott saying that for the first time in over two years a swell was felt on the ship. Much to their delight, the ship was working in her icy bed and rising a foot now and then. This was good news indeed, but eight miles was still a long way, and with the very limited season it was a large area to break away.

We were by this time south of the glacier snout off the Erebus shore, the glacier being ten miles from Hut Point. The *Morning* ran alongside the glacier one day, and took on board the coal left there the previous year. The *Discovery* had only taken a few tons from the stock.

As the ice broke away from the sea edge, so the half-way camp had to be shifted south to equalise the distance between *Discovery* and relief ships. The sledge journeys became proportionately quicker, though bad weather caused repeated delays.

STILL SIX MILES

On February 3 we were six miles from Hut Point. Captain Scott was naturally very anxious, and thought the ice was going out too slowly. The season was advancing rapidly, and although the temperatures were good they might fall at any time and start freezing the sea again.

The *Discovery* was being gradually dismantled. All valuables were being sledged over to our ships. Her men were very low-spirited at the prospects of having to leave their good old ship, and were hoping against hope that she would be freed at the last minute. We received most pathetic letters from them, per dog-team mail service, recently inaugurated. We kept them posted in our doings at the outer edge, reporting every hundred yards that broke away.

On February 5 blasting operations were carried out on an extensive scale. Captain Scott started them going himself. The ice was now from five to six feet thick, and very solid indeed to dig holes through. Holes were dug in a direct line towards Hut Point, at intervals of sixty to eighty yards. All kinds of implements were improvised for this work.

THE VOYAGES OF THE 'MORNING'

Owing to the cold the picks and crowbars became brittle and frequently broke. However, it was astonishing what ingenuity prevailed, and bars were made out of any odd pieces of iron and steel and pointed in some serviceable way. The hole-digging parties suffered much from snow blindness in the incessant glare. A hole usually took from three to four hours to dig. At times this work was carried on in reliefs for a day or two at a stretch.

Captain Scott was with us until February 9, when he returned to the *Discovery*, He then started blasting round his own ship, and she appeared free, with the exception of the propeller aperture. The propeller was in a solid block ten or twelve feet thick, and it was risky to blast too closely in case of damaging that vital part of the ship.

A few anxious days now passed. Very little ice broke away, and the temperatures were falling. The hole-party worked solidly, and soon had holes ready for over one mile in.

On February 11 Captain Scott issued our sailing orders. The *Morning* was to sail for New

THE DARKEST HOUR

Zealand on 27th inst., taking the scientists. The *Terra Nova*, being the more powerful steamer, was to remain until the second week in March. If the *Discovery* were not freed by then she would be abandoned, and the remainder of her party would return in the *Terra Nova*. These orders were felt alike by all to be almost a death-warrant. If Nature would only come to our rescue now! We could do nothing without her assistance. Oddly enough, on this very afternoon a change occurred. Long wisps of cirrus clouds hung over the top of Mount Erebus. We assumed that a gale was blowing on the other side in the Ross Sea, and sure enough, in a few hours, a long grateful swell came down the sound. The blasting was continued now with alacrity, and large floes broke away with every explosion. It was a thrilling sight. About 6 P.M. I went over with a note from our captain to Captain Scott, informing him of the good news, and requesting him to send all available hands to assist. A 'lake' of open water extended off Hut Point, caused no doubt by thaw, over a known two-fathom shoal off the point. I had consequently to climb over the point and down the other side to the winter harbour.

Being the herald of such good tidings, I was greeted with cheers, and Captain Scott immediately despatched Royds with a party. They worked all night. At midnight I went up to Hut Point, and could hardly believe my eyes. The ships were not three miles off! I returned to the ship next day, and found that the swell had almost stopped. The hole-party, however, worked away with redoubled energy, in anticipation of the next swell, which surely would be the grand finale! The ice broke up so rapidly the night before that the half-way camp went out to sea on a floe! The *Terra Nova* steamed after it and rescued it with some difficulty.

Captain Scott came out again, and was highly satisfied with the fine break-up. Naturally, we were all rather excited by this time, but still wondered if our efforts would be successful. Whilst a swell prevailed, hopes were high, but as soon as it subsided things assumed a different aspect. Two miles of ice were a mere nothing to

AN OCCASIONAL INCIDENT—ADRIFT ON THE SEA-ICE

SUSPENSE

eighteen; but two miles is, nevertheless, a terribly long way in a falling season. Only a quarter of a mile broke away during the next two days. If the temperatures were to fall suddenly our chance was lost. This was quite possible. At this time during the previous season minus temperatures were of daily occurrence, and the sea had started freezing over the sound. The suspense was horrible, and without Nature's assistance we were helpless.

Sunday, February 14, was boisterous. A strong, biting, easterly wind blew, accompanied by heavy squalls with driving snow-drift. It was unpleasant working on the ice, but time was limited now, and discomforts must be faced. Parties worked all day digging holes, and several successful blasts were made. The large floes soon broke away completely and drifted to the north-west.

As the afternoon wore on the weather moderated, and the ice was noticeably thinner. At 5.30 P.M. a wonderful thing happened. The entire ice field suddenly cracked in all directions as far as we could see. It was eerie in its suddenness. The great work of the moment was to rescue the men and gear. They were hailed to return on board at once. This was attended with much difficulty, owing to the floes breaking up into such small pieces and drifting away rapidly. The men were hauled on board just in time, but a lot of gun-cotton and tools were lost. It was rather pathetic to see the last few holes, laboured at in the blizzard, drifting out to sea in the floes! The ice rapidly decreased in thickness from four feet to two. At 6 P.M. the wind fell light and the sun shone brightly. Both ships now worked with much vigour, charging between the floes to force a way through. Back we went, astern for some distance, and then would come up with a full-speed rush, crashing desperately into the rotting floes and pushing them apart. And so it went on for some hours—astern and ahead—with grim determination, each charge gaining on the last.

The excitement was intense. It was now certain that we could reach the winter harbour, and it resolved itself into a sporting struggle for priority. The *Terra Nova* was the more

THE LAST RACE

powerful, but by strategy we kept them on the alert. Somehow we got as far ahead at each charge. The *Discovery's* people had assembled on Hut Point—now not half a mile away!—and witnessed this great and final race for our goal. The *Terra Nova* taught us a useful wrinkle. We noticed that she rolled heavily owing to her crew running from side to side at given signals. We quickly realised the advantage of this in working the ship, corkscrew fashion, through the ice. We therefore adopted this plan ourselves, and very shortly both ships were rolling violently and charging away at the same time. We shouted ourselves hoarse with excitement as we rushed from side to side of the ship. I can safely state that this is the most thrilling and sensational experience in my life. At 10 P.M. the open water off Hut Point was but one hundred yards off! We had been at a great tension for some hours now, but we were not tired. Rest could be had another time. This struggle must be fought through, now or never.

At 10.30 P.M. the final crack was made,

and the *Terra Nova* had the good luck to be charging ahead. The little *Morning* stuck to it doggedly, but her chance was missed by a few seconds. The *Terra Nova* pushed in at full speed, and as the floes parted she gradually gathered way and steamed victoriously into the open water off the point. The *Morning*, steaming in the wake, arrived immediately afterwards.

As the ships passed Hut Point a grand rousing British cheer broke from the gallant band on the point, and as we responded, with a cheer that shook our very frames with an indescribable thrill, a Union Jack fluttered out bravely from the staff on Hut Point.

CHAPTER XIV

Preparations for freeing *Discovery*—*Morning* adrift—Terrific explosions—*Discovery* released—Thrilling scene—A severe blizzard—*Discovery* driven ashore—Heavy weather—Trying experiences—Ships alongside glacier snout—Transhipping coal and stores—' Follow me '—Start of homeward voyage.

THE dramatic entry of the relief ships into the winter harbour was deeply impressive. For two years the *Discovery* had been held in this secluded corner, and even to within a day or two the odds were against her ever being free to sail the seas again. The more we reflected upon our wonderful good luck, the more we realised how dependent we were at every turn on Nature's assistance. Gun-cotton, no doubt, helped materially; but Nature was the potent factor.

At Captain Scott's first lecture in the Royal Albert Hall, London, the picture thrown on the screen, of the three ships,

THE VOYAGES OF THE 'MORNING'

secured to each other in the winter harbour, created a thrilling acclamation, and our hearts thumped with pardonable pride, as we mentally lived over again those arduous and anxious days, spent in achieving this historic work.

But we are still in the Antarctic, and many varied experiences must be passed before we are privileged to attend that memorable function at the Royal Albert Hall. As soon as the ships were fast to the solid ice of the winter harbour, the cheering was renewed. The *Discovery* men rushed down from the point, and much hand-shaking and many congratulations were exchanged. As the total muster of men from the ships numbered one hundred, the scene was most animated. A bumper gathering was held on the *Discovery*, which could hardly be otherwise than one of delightful *camaraderie*.

A few heavy floes broke away from the harbour, but the *Discovery* was still encased in her solid bed. The fine weather only lasted for a few hours. A strong southerly wind arose, and the *Morning*, levering out

THE 'MORNING' ADRIFT

her ice anchor, was unfortunately driven out of the harbour. I was on watch, and, fearing that the ship would drift on to the shoal off Hut Point, I immediately rang down for the engines. A great deal of puffing and snorting took place in the engine-room, and a corresponding hissing of steam exhausted through the steam-pipe, but with no practical result. As we were drifting dangerously close to the point, I rushed below, and anxiously sought the stoker-engineer in charge. As I slid down the little ladder, he cheerfully greeted me with, 'It'll be all right in a few minutes, sir; the fact is, she's a bit stubborn to-night!' Fortunately we scraped past the danger, and in the few minutes, which seemed like hours, the engines struggled out a few revolutions. The strong wind and drifting ice embarrassed the ship so much, that three hours passed before she was securely fast in the harbour again.

Many hands from the three ships worked all day and through the night, sledging loads of ice, for thawing, and filling the *Discovery's* boilers. Others got the *Discovery*

THE VOYAGES OF THE 'MORNING'

ready for sea. Much had to be done, naturally, to a ship laid up for two years. One party dug a large hole some yards ahead of the *Discovery's* bow, in preparation for a blast. The ice, being ten feet thick, made this a difficult undertaking.

During the afternoon, a heavy gust of wind blew the unfortunate *Morning* away again, and as the engines were still ' stubborn,' she collided with the *Terra Nova*. The *Terra Nova's* bowsprit shot through our port main rigging, and the side of our bridge was smashed. It was exceedingly difficult to secure the ship firmly to the ice. The *Terra Nova* was in a better berth, and had a wire fast to the *Discovery*. We were very happy, though, in spite of these occasional troubles, and many visits were exchanged between the ships during the stay in Winter Harbour.

Early next morning Captain Scott set off an alarming explosion of four charges of gun-cotton, simultaneously, in the hole ahead of the *Discovery*. The report was terrifying, and as the ships were so much shaken, fears were entertained for the *Dis-*

THE FINAL BLAST

covery. As far as we could see, however, she appeared intact, and as the explosion caused a great rent in the solid bay ice, the result was satisfactory.

The *Discovery* was now ready for sea. All gear, in the shape of sledges, stores from the hut, &c., was shipped; sails were bent, and running rigging put in working order. The final blast was to be made in a hole in a tide-crack well on her port quarter. The charge was placed, and the circuit wire led to the battery on the stern of the ship. The large gangway was drawn on board, and the ship's company mustered on the after deck.

Just before noon the charge was exploded by Captain Scott himself. The concussion was severe, and the ships were much shaken. In a few minutes, the *Discovery's* stern, which had been weighed down by the solid ice surrounding the propeller, suddenly rose to even keel, and immediately afterwards the whole surface of the bay ice cracked —like streaks of forked lightning—in all directions. The solid, irregular-shaped floes then slowly drifted out of the harbour, and,

THE VOYAGES OF THE 'MORNING'

thrilled with a supreme joy, we witnessed at last the release of the good ship *Discovery* from her icy prison.

Freed from all ice, she slowly swung round with the wind to her anchors, and a stirring cheer rang out from her loyal consorts. Ensigns were dipped, and whistles hooted, and in our joyous abandon we knew that we looked upon a scene which would ever be memorable in the annals of the Antarctic.

The bay was soon clear of ice, and the *Terra Nova* made fast alongside the *Discovery*, preparatory to coaling her. The *Morning* cruised off the harbour. Captain Scott signalled us to anchor near them, but the *Morning's* windlass was so old-fashioned, and took so long to heave up the anchors, that our captain thought it safer, in the event of a sudden blizzard, to remain under way.

That evening the three captains were to dine together on the *Discovery*, this being their first meeting on that ship. I took our captain over in a boat about 6 P.M. It was getting thick with falling snow, but the wind was light. We decided to steam

A STRUGGLE WITH A STORM

up to the ice face, now to the southward of the winter harbour, and return for the captain at midnight. However, our anticipations did not come off.

By 10 P.M. it was very thick, and the wind moaned ominously in increasing heavy gusts. This wind, oddly enough, came from the westward, quite an unusual quarter, and forced the ship through the rotten slushy ice towards Cape Armitage—the southern point of the winter harbour. The ship rapidly gathered way, and no efforts on the part of the feeble little engine could pull her up.

As we were now making directly for Cape Armitage, we set some sail, in the hope of weathering it to the northward, and getting away up the sound. The slushy ice, bearing on our side, upset this manœuvre, and the only course open to us now was to take a risk, and run to the southward of the cape. It was very thick, with hard driving snow, but so closely did we round the cape, that its bold outline was seen hazily through the drift. Our main topsail was rent in several places, and was nearly blown out of its bolt-ropes. All canvas was taken in, and as bad luck would have it, the wind suddenly shifted south, and increased to a furious blizzard.

We were in a very awkward predicament, as the cape was now immediately to leeward, or right astern. The engines had to be relied on solely. The boilers had given much trouble lately owing to defective tubes, and we hoped and prayed that they would hold out until the gale abated. The squalls were violent and shrieked in their fury, each one driving the ship astern, closer and ever closer to the rocky cape behind us. The gale raged all night, and at times the ship's stern almost touched the rocks. Ice broke away ahead all the while, which assisted to retard headway.

Providentially, the gale abated as suddenly as it arose, and at 6 A.M. we rounded the cape, and steamed to the winter harbour. I then returned in the boat for our captain. The *Discovery* rode out the gale at anchor, but the *Terra Nova*, after carrying away several wires and bits of rail, cast off about 10 P.M. and weathered the gale at the ice-face, a few miles away from us.

LEAVING WINTER HARBOUR

The three-captain dinner-party was a failure. Much anxiety for our welfare was felt on the *Discovery* during the night, and our captain was greatly relieved when I reported that all was well, although we had spent a very trying night.

Captain Scott then decided that the three ships should proceed to the glacier snout, ten miles north, to tranship the necessary coal and provisions. This, being more like a wharf, was better than lying at anchor or cruising about. Three nights of more or less continuous work and anxieties were beginning to tell on us, and we longed to get away from this stormy locality. The *Discovery* had to be coaled and provisioned, and the sooner this was done the better for all concerned.

Soon after noon the *Discovery* hove up her anchors and got under way for the first time since her imprisonment. The weather appeared treacherous, and just at this time an ominous lull prevailed.

We were lying a few miles away from Winter Harbour, but the *Terra Nova* was lost in the misty south; indeed, she had not been seen since leaving the *Discovery* the night before.

The *Discovery* had only just put her 'nose' out of the harbour when a sudden squall broke from the southward. The ship, being very light, presented a high side to the wind, which, striking her broadside on, drove her bodily to leeward. Unfortunately she had not gathered enough headway to haul up to the windward, and was rapidly driven on to the shoal ground off Hut Point. We witnessed this calamity with dismay. She took the ground, heeled over alarmingly, and remained fast, whilst the merciless icy seas dashed over her. In a few minutes the dense drift blotted out the entire surroundings. We were naturally much concerned, and wondered if, after all our labours, it was the decree of Providence to hold the *Discovery* in the Antarctic for all time.

But our helplessness to render assistance was the most galling feature. The blizzard increased to such a fury that the *Morning* was literally blown away across the sound, and, labouring heavily in a short breaking sea, she was frequently smothered in sheets

THE 'DISCOVERY' AGROUND

of icy spray. We could do nothing for the *Discovery*, and the storm was so violent that we soon realised that the utmost skill was required for the safety of our own vessel. We hoped that the *Terra Nova* would, somehow, find out the *Discovery's* misfortune, and, at any rate, stand by her in case of a disaster. However, as she was not in the vicinity at the time, it was improbable that she could do this.

It was incredible to think that neither relief ship could be of avail in this predicament. The affair was a nightmare to us. No one, sitting in the comfort of an armchair before a cheerful fire at home, reading about—as we feared—the loss of the *Discovery*, would have any sympathy with the crews of the relief ships for not going to the aid of the stranded vessel. We should probably have been branded with neglect or incompetency. But an Antarctic storm must be experienced to realise how utterly impossible it was to do this, especially in ships of limited power.

The blizzard raged fiercely all night. Eyes were bloodshot and sore, and faces smarted with the stinging spray and cutting icy particles that drove before the wind. Under such circumstances sleep came but fitfully, and we wondered how much longer we could stand this severe strain.

By midnight the ice-hummocked fore-shore on the western side of the sound was sighted, and the ship was immediately put about. We had only a vague idea where we were, but it was certain that we were well to the north end of the sound.

The peculiarity of Antarctic weather is the suddenness of its changes. Early the next morning the wind abated rapidly, the drift ceased, and the overhanging pall of a few hours before vanished. We were abreast of Beaufort Island, and had consequently been blown about forty miles away. We steered in the direction of the glacier snout on the eastern shore. Some hours passed before the familiar landmarks were distinguished, and it was late in the afternoon when we sighted, to our intense relief and joy, six masts together at the glacier. We cheered with delight, knowing now that, whatever may have happened during the past awful night, the two ships were safe.

It is strange how quickly anxieties and

SNATCHED FROM DESTRUCTION

hardships are turned aside. We were weary, and possibly haggard, but we were nevertheless perfectly happy. The weather was improving, and were not the two ships only a few miles distant, apparently all right ?

In the evening we arrived at the glacier, and according to custom now, we cheered the ships vigorously. The *Terra Nova* was alongside the *Discovery* discharging coal and stores. The scene on the ships was very animated, and every available man was busily employed, including the scientists. The learned and respected bacteriologist and the biologist, attired in the most mundane rig-out, with sou'westers on, were seriously dragging bags of coal along the deck, and filling the bunkers. We soon joined in the common toil.

We then learned of the *Discovery's* mishap. She bumped heavily on the shoal for several hours, and broke away part of her keel. At one time, after every endeavour to get her off the rocks had proved unavailing, it was considered a hopeless case, and her gallant but distressed crew awaited her destruction. Owing to some remarkable change in tide or current, the stern swung into deeper water.

THE VOYAGES OF THE 'MORNING'

The engines were soon away again, and by rolling the ship continuously, she gradually gathered sternway, and came off the dangerous shoal. She passed the rest of the night in the howling gale at the ice face. With the clearing weather next morning the *Terra Nova* was sighted close by, and they proceeded in company to the glacier.

By midnight all available coal and stores were transhipped from the *Terra Nova*, and the *Discovery* then hauled alongside the *Morning*. We had only a small amount of coal in our bunkers, but as the *Discovery* had less, and needed as much as possible to enable her to carry out her cruising programme before finally returning to New Zealand, we gave her really more than we could spare. The coal was worked out as expeditiously as our tired limbs would permit.

This was now the fifth arduous night in succession, and the strain was telling. Evans, the doctor, and I worked at the coal whips, hauling up the bags by hand. We competed with three seamen, in spells of half an hour. In spite of our exhausted condition, we worked cheerily, and held the

'FOLLOW ME'

record by pulling up eighty-seven bags in one half-hour spell, to the men's eighty-five.

At 6 A.M. the work was finished, and the *Discovery* shifted along the glacier for watering. Thoroughly played out, we then enjoyed a much-needed rest, and slept for six grateful hours on end. We could well have slumbered longer, but another gale set in. A short nasty sea made, causing the ship to bump heavily against the glacier. As the rudder and propeller were receiving dangerous jars against the projecting ice spurs, all hands were called, and after an unavoidably risky manœuvre, we backed away from the glacier. The *Terra Nova* hauled off earlier, but the *Discovery* was still ' watering.'

At last, at 5 P.M., all work was concluded, and the *Discovery* cast off from the glacier. With an escort on each side, she got under way, and hoisted the signal : ' Follow me.' With squared yards and bellying sails, the ships ran bravely before the gale to the north ; and with memories of joys, hardships, and anxieties, we repeatedly looked back, fascinated, at the receding landmarks, now viewed, for all we knew, for the last time.

CHAPTER XV

A beautiful Antarctic scene—Ships in company—Impressive coast-line—*Morning* parts company—*Terra Nova* escorts *Discovery*—Boisterous Southern Ocean—Engines break down—An anxious month—Reverses making Auckland Island—Arrival Port Ross—A peaceful contrast—*Discovery* and *Terra Nova* at rendezvous—Pleasant days—Ballasting ship—News of outer world—Departure from Auckland Island—A fine slant—Return of National Antarctic Expedition.

THE gale gradually decreased as the ships sped north. By midnight McMurdo Sound was left behind, and all drift had disappeared. The night was beautiful. As one walked aft on the little poop-deck the wonderful scene to the southward was magnificent, striking enough indeed to throw an artist into an ecstasy.

To the left towered Mount Erebus, whose icy slopes were alternated in dazzling sun-touched ridges and shadows of the deepest purple. The ice-capped Beaufort Island lay

A BEAUTIFUL ANTARCTIC SCENE

at the foot of the mountain. To the right ranged the lofty pink-tinted peaks of the western mountains along the mainland, interveined by massive glaciers. The mountains gradually deepened in shade as they appeared in succession to the south, and in the miraged distance the further peaks stood out boldly against the fiery glare of the dipping sun. The southern horizon was ablaze, in shafts of gold and crimson, which tinted the face of the innumerable icy slopes. In picturesque contrast were the dark shades of purple and grey on their northern sides. The occasional bergs and drift ice floated listlessly, and stood out strikingly in the indigo-coloured sea.

But it is presumption to attempt to pen-picture this last impression of McMurdo Sound, for it was indescribably beautiful. However, my feeble effort may tend to show that, although we had suffered many hardships and anxieties during the last week, the fascination of the place still remained, and, with a feeling akin to sadness, I reluctantly turned my head away from its awful charm.

THE VOYAGES OF THE 'MORNING'

The three ships in company steered to the north along the western shore, which comprises a continuous chain of ranges, very similar to the mountains of McMurdo Sound. Some of the glaciers were striking and extended well out to sea. One in particular jutted out for several miles and could well be termed a barrier. The ships had to steer to the eastward for some distance to round its tabular-shaped extremity.

Still carrying the southerly wind, Mount Melbourne was passed the following night. The seaward side of this wonderfully symmetrical cone terminates in Cape Washington, the southern point of Wood Bay.

The *Discovery* required as escort along the coast either one of the two vessels. We should have appreciated cruising with her very much, but it was quite evident that with our weak-powered engines we should only have delayed her in her work. As she was going into Wood Bay to complete her magnetic work, and the favourable breeze still held, our captain signalled for permission to proceed to the prearranged rendezvous—Port Ross, Auckland Islands.

PARTING COMPANY

The satisfactory reply soon came, viz.: 'Proceed on your voyage.'

We parted company, therefore, off Wood Bay, and the *Terra Nova* was retained as escort. Yards were squared and, standing to the north-east, the *Morning* bowled merrily on her way.

It soon set in gloomy, with frequent mists of driving snow, and a very keen watch was necessary to avoid the dangerous stray bergs. In five days' time no more ice was seen, and we became deep-sea sailors again. The days shortened rapidly, and, after the strain of two months' incessant daylight and trying glare, night proper was thoroughly appreciated, and the first stars seen were welcomed with a cheer.

The Southern Ocean was particularly boisterous during the northward voyage. Rarely is fine weather experienced there, but the month of March is perhaps the stormiest. However, I think that our experience of continuous gales from north-west and west for a period of twenty-four days on end, without cessation, just about constitutes a record. It was extremely tantalising, and as the Auckland Islands lay to the north-

THE VOYAGES OF THE 'MORNING'

west, or to windward, great patience and dogged determination had to be exercised. It was a trying time for all. The engines would have assisted materially in keeping the ship from making excessive leeway, but they had unfortunately given out, and were under repair for some weeks.

But perhaps the most disturbing feature was the lightness of the vessel. There was practically no coal in the ship, stores were diminished, and the only actual weight was the engine. Hitherto the ship had been well ballasted, but we could not afford to take any liberties now. The seas were mountainous, and the ship lurched dangerously. How we longed for a fine day with a smooth sea! It never came, so we had to make the best of a rough-and-tumble existence. It was a hard month, and it told on us. For the average person it would be impossible to imagine the discomforts attached to such conditions in a little tub of a vessel of 290 tons. We were cooped up the whole time. Exercise came only by the frequent trips aloft to furl or loose sail, for the decks were but a dreary con-

BAFFLED OFF THE AUCKLANDS

tinuation of the turbulent ocean. During one heavy gale another boat was washed away and the bridge damaged.

When we did eventually get within sighting distance of the island it set in thick, and a strong gale blew us away before it cleared. The day afterwards we beat back again, and during a temporary clear in the evening a light was seen for a few minutes. This we surmised to be from the *Discovery* or *Terra Nova*.

On the following morning the gale abated, and blue sky greeted us for the first time. The engines were now sufficiently repaired to enable us to proceed slowly. All sails were clewed up, and we steamed at the depressing rate of one knot towards the supposed direction of Auckland Island.

The land was at last sighted, but we were too late to make the harbour before nightfall. Arriving off the entrance of Port Ross at 8 P.M., we dodged about a few miles off the shore, intending to enter at daylight. To our intense disappointment the breeze freshened during the night, and blew us so far away again that at daylight no land was to be seen. This reverse was an exasperating addition to our many misfortunes. However, a last supreme effort was made. In desperation, the chief engineer risked a few more revolutions out of his delicate charge, and the baffled, weather-beaten *Morning* now struggled bravely at two knots. By great good luck there was a 'set' in to the shore, and the wind and sea decreasing as the land was approached, we reached the entrance that afternoon. This was Sunday, March 20. McMurdo Sound had been left behind on February 20—truly a weary month of wretched discomfort, anxiety and hardship.

Oh, the joy of it, as the little ship steamed between Enderby and Ewing Islands, and into this haven of rest, happily named Sarah's Bosom! As soon as we rounded into the harbour the masts of a vessel were seen over a point in the upper reach, so we anchored in Erebus Cove, about five miles from the entrance.

The sudden contrast between the present peace and the stormy stress of the past months was great. Here at last the ship lay

REUNION

quietly. The placid water sparkled in the sunlight, and the dense luxuriant vegetation, growing to the very water's edge, continued in reflection in imaginary depths. The landlocked harbour, dotted with thickly wooded islets, assumed more the appearance of a beautiful lake. But it was the greenery that was so particularly pleasing. After the strain of the eternal Antarctic glare, our eyes simply feasted on this, and the peaceful environment amply compensated for our recent discomforts.

A boat was soon lowered, and a party pulled round to the inner harbour. The *Terra Nova* was anchored there indeed, and further in—O happy day!—was the *Discovery* also. So here we all were once again, in joyous company.

The *Terra Nova* and *Discovery* parted in a blizzard soon after we left them, yet, in spite of the separation, the three ships arrived at the rendezvous within a few days of each other.

Evans and I landed on Pig Point close by and took a series of observations to verify our chronometer rates. We then visited the provision depot for the shipwrecked at Erebus Cove. There are many similar depots maintained by the New Zealand Government at the outlying dangerous islands round New Zealand. The hut is small, but a haven indeed for castaways. In one part are stores and clothing and a fireplace; in another some rude bunks. On the walls are cut the names of many men who have been cast away there, and there are several memorials to crews who have lost their lives, either from drowning or starvation. Near the hut there is a peaceful little cemetery, where rough wooden crosses have been placed over the graves. Several are marked 'Unknown,' one being an infant aged five months. It is a sad spot. Before leaving the island we added some stores and tobacco to the stock in the hut.

A party from the *Discovery* visited us, and we had an interesting evening relating our recent trying experiences. The three ships had had a baffling time with the elements, but it was evident that the *Morning* had fared the worst.

Next evening the *Morning* officers dined on the *Discovery*. A fair wind enabled us to sail over. The ward-room was artistically decorated with huge tree ferns and the table

BALLASTING

tastefully arranged. To give a little tone, the menu was typed in French. It was the happiest dinner party imaginable, but the return 'home' against a strong head wind rather took away from the enjoyment. In the intense darkness we pulled mechanically for nearly two hours.

Owing to our vessel's unballasted condition, we took in ninety tons of large round stones, with which the foreshore was strewn. This occupied some days in the boats. We usually started at 6 A.M., and the first boat to go in generally disturbed a stray sea-lion or two. They barked viciously at our intrusion, and, in apparent disgust, waddled away into the dense scrub.

The New Zealand Government steamer *Hinemoa* visited Port Ross during our stay, and we learned the news of the outer world. We heard then of the Russo-Japanese war. The newspapers were particularly acceptable and eagerly perused.

To commemorate the anniversary of our arrival at Lyttelton from the Antarctic, viz. March 25, 1903, we had an original dinner of the products of the island. These comprised mussels, fricasseed cormorant, sea-lion steak, and mutton-bird. Fresh water there was, too, from the clear springs of the place, but a little champagne was also introduced to make the event official.

A day or two before we left Evans and I spent an afternoon ashore washing clothes in the clear fresh stream running into Erebus Cove. Washing clothes had become an accustomed necessity, but we had never enjoyed such an unlimited supply of water. We sat on logs thrown across the creek, dangling our sea-booted legs in the stream, and with pipes alight, we started off merrily. But how akin is pleasure to pain! In a few minutes we were severely attacked by the most persistent sandflies, and faces, necks, and bared arms suffered unmercifully. As we could not possibly refrain from rubbing the smarting stings, our faces, indeed our very heads, were soon smothered in soapsuds, which produced a ridiculous effect. We braved out the attack, but did not forget that afternoon in a hurry.

Early on the afternoon of Tuesday, March 29, the ships steamed out of Port Ross.

RETURN TO LYTTELTON

A fine strong S.W. wind blew, and, with all sail set, we bowled along at eight to nine knots towards New Zealand, keeping within a few miles of one another.

Stewart Island was sighted on the following forenoon, and at midnight the lights of Dunedin could be seen. The wind decreased gradually, and the weather remained very fine. It was calm next night as we drew up to Banks Peninsula, and the little ships stood out darkly on the silvery moonlit sea.

At the first streak of a beautiful dawn —April 1, 1904—the ships arrived off the entrance to Lyttelton Harbour. As the pilot tug steamed towards the little fleet ringing cheers rent the still morning air. We passed through the Heads in file, *Terra Nova*, *Discovery*, *Morning*, and slowly steamed up the harbour. It was Good Friday, and the news travelled quickly.

Before the ships were half-way in innumerable craft of every description were cruising round us in figure-of-eight manœuvres. Flags were flying, whistles tooted, and cheer upon cheer rang out from those who had risen so early to welcome us home.

THE VOYAGES OF THE 'MORNING'

Lyttelton was gay with excitement. Flags flew from public buildings, from private houses, and from every ship in the port. The tug, which had quickly returned from the Heads, now approached, crowded with our good and loyal friends. Evans excitedly pointed out my fiancée, and I his. A band played 'Home, Sweet Home.' Oh! it was a thrilling time.

As the ships drew up to their respective berths a continuous cheering broke from the packed hundreds on the wharves, handkerchiefs fluttered distractingly, shrilly blew the whistles, and loudly boomed the guns.

A triumphant entry indeed, and a glorious reception. Thus the National Antarctic Expedition had returned.

CHAPTER XVI

A great welcome—Evans's wedding—Departure for England—Around Cape Horn—The Falkland Islands—Last run of voyage—Depressing head winds—Change off Cape Finisterre—Running before a gale—The dear Homeland—*Morning*, steamer or sailer?—Plymouth Sound—The end of the cruise.

IF it were not for contrast life would be exceedingly dull, and in our adventurous expedition we certainly experienced its extremes. The hospitable welcome and generous entertainment extended to us from all quarters formed indeed a delightful change from the wretched discomfort and anxieties of the past months. Enough cannot be said of the great kindness shown to every one of us by the good people of New Zealand.

By way of compensation for our recent trials, the captain granted the entire crew one week's leave. This thoughtfulness was much appreciated and strengthened the loyalty of our hard-working men. The ship was therefore 'closed' and handed over to a ship-keeper.

A week or two after our return a social event took place of particular interest to the Expedition and its numerous friends, namely, the marriage of Evans to Miss Hilda Russell. It was a very happy occasion, and, naturally, I was my old chum's best man.

The little church was prettily decorated, and the *Morning* sailors were in attendance as a sort of guard of honour. Cap ribbons marked *Discovery*, *Morning*, and *Terra Nova* streamed from the bride's and bridesmaids' bouquets. The bridegroom and groomsmen wore uniform. A large number of guests attended the reception afterwards, and congratulated the happy pair.

As usual at such functions, there were several amusing incidents. There is a custom at naval weddings for the bride to make the first cut in the cake with her husband's sword. This rite was duly performed, but, by an oversight, the blade was still coated with vaseline. The *Morning* sailors were each given a glass of champagne to drink the health of the bride's father. The bosun made a

LIGHTS AND SHADOWS

characteristic speech, and down went the wine. One man, nudging his neighbour, said,—' By gum, Jim, but that's ruddy fine ginger ale ! '

But how transitory is life. On the ninth anniversary of their marriage my dear chum's wife was taken from him after a brief illness. This great sorrow befell him, sadly enough, just as he had attained distinction and honour at the conclusion of Captain Scott's last fatal expedition.

After the necessary docking and overhauling, the ships left New Zealand early in June for England. For many this was a final parting from numerous good and kind friends. It was a sad wrench. We sailed away across the Southern Ocean, and, despite the winter season, experienced favourable breezes and moderately fine weather. Towards the end of July the dreaded Cape Horn was rounded on a peaceful starlit night, and in a day or two we put in to Port Stanley, Falkland Islands, for coal. The *Discovery* was just leaving when the *Morning* arrived. The ships had parted company the day after leaving New Zealand, and it was a coincidence that they should again meet at the Falkland Islands.

I do not know what this part of the British Empire is like in the summer, but in winter it is a dreary spot. The township was under snow, and the frequency and violence of the gales were depressing. Owing to repeated bad weather, we could only lie alongside the coal hulk on two occasions in the week we were there. Ninety tons were taken in, and repairs effected to the almost worn-out boilers. The inhabitants of this desolate little colony showed us some hospitality in the shape of occasional modest entertainments.

However, we were glad to heave up the anchor and enter on the last run of our voyage. A fine 'slant' enabled us to reach the equator in a few weeks' time. The engines were only used occasionally, but, much as we nursed them, they unfortunately broke down again when most needed to push against the north-east trade winds. The 'trades' blew well from the north at this season, and we experienced them from

WEARY WINDS AND A PROPHECY

the equator until well north of the Azores. This meant incessant tacking, and the ship beat so poorly to windward that we thought we should never reach England. We were naturally very keen to get along quickly at this time, and this reverse was tantalising. The weather soon became the chief topic of conversation, and each day caused our estimated arrival date at Plymouth to be put ahead further and ever further with disappointing consistency.

Meteorology was discussed at each meal, and many futile weather forecasts were made. The bosun's suggestion one afternoon, although it likewise did not come off, displayed, at any rate, an original terseness. The sky was cloudless and the sun just setting. He was keeping part of the chief officer's watch. The captain, passing along the poop on his way to dinner, asked him his opinion of the weather. Smugly and deferentially came the reply: 'Well, you know the old saying, sir—

> When the sun sets as clear as a bell
> Easterly wind as sure as——!

Well, you know what I mean, sir.'

But he was no prophet either. The

northerly wind still held persistently. Four months had elapsed since we left Lyttelton, and the voyage was becoming monotonous. All blacking down, painting, and the numerous homeward-bound titivations were finished, and the *Morning's* smart appearance threatened to wear off if the journey were much more protracted. But a north wind cannot last for ever. About the middle of October, when abreast of Cape Finisterre, a south-west breeze was gladly welcomed. Yards were at last squared in off the backstays, and as the wind freshened so we gathered way. It set in thick with driving rain squalls, but we bowled along in fine style, keeping a vigilant look-out in doubled watches. The wind held steadily in direction, and gradually increased to a hard gale. The little ship heeled to the strain. After the recent baffling winds we were keen to take every advantage of the following gale. As the result of the good progress made, the ship was in the mouth of the English Channel on the evening of the third day. Soon after midnight the wind suddenly shifted in a heavy squall to north-west, the misty rain

UNDER STEAM OR SAIL ?

ceased, and in the crisp, clear, starlit night, the Scilly light was sighted to the northward. Evans sent below for me, and we gazed with joy at this first glimpse of the dear Homeland after an absence of two and a quarter years.

An odd thing happened that night. After much worry and toil, our skilful engineer had patched up the shaky old engine. This was started, amid cheers, for the final flutter down 'the straight,' and, accelerated by the full spread of canvas in the strong breeze, the ship was going her utmost speed. Being in the track of shipping, a masthead light was placed in addition to the side-lights to indicate that we were a steam vessel. The ship drove so fast before the breeze, however, that the propeller could only be sportingly termed as an 'also ran.' It performed many more revolutions than the feeble engine could possibly make unaided, and it seemed that the way of the ship was turning the engine, and not the engine causing the way ! Perhaps an occult method of ship's propulsion may evolve from this observation at some future date. But this is beside the incident.

THE VOYAGES OF THE 'MORNING'

A steamer bore down on us, and, according to the Rule of the Road, it was our duty to keep out of the way. Seeing that to do this effectively we should be 'caught by the lee,' the line of least resistance was adopted, viz. the masthead light was doused, and the ship proceeded as a sailing-vessel! The Board of Trade might not appreciate this, but in the circumstances it was the safest plan for all concerned.

Next day we signalled our number on passing the Lizard Lighthouse, and, the sea becoming smoother, we sailed peacefully over the last few miles along the beautiful shores of Cornwall. A cruiser homeward bound from China, flying the pay-off pennant, steamed closely past us, and her blue-jackets manned the rigging and gave us three rousing cheers.

Late in the afternoon we steamed into Plymouth Sound. Two men-of-war cheered us vigorously as we entered, and their bands played inspiring airs.

As night closed in the anchor was dropped, and signalised at last the termination of the *Morning's* adventurous cruise. England

HOME

and home! Our mission was over. The end seemed to come with a strange suddenness. As one leaned over the little ship's rail in the peaceful night and gazed at the glare of the lighted town, with the dancing reflection of innumerable lights in the darkening water, feelings of joy and sorrow were intermingled. There was a feeling of pride, too, in the reflection that we had done our best in the Expedition's cause and, despite discomforts and reverses, had done our duty in the arduous undertaking. The prospects of being once more with our dear ones, and to be able so soon to gather round the homely fireside and relate our thrilling story, was a joyous thought. But sad was the anticipation of the breaking up of our staunch little company, who for over two years had lived and toiled so cheerfully together. A bond of affection had been established, and, although we must now disperse and go on our respective ways, there was some consolation in knowing that the memory of our happy comradeship could never be erased.

CHAPTER XVII

Morning laid up at Devonport—Our diminutive floating home—Dividing the 'spoils'—Crew paid off—Sailors' affection for their ship—The last of the *Morning*—Homeward-bound crew—Our unconscious humorist—The Great City—Home, sweet home.

ON the following day, the *Morning* was the centre of much interest. Visits were paid by the Admiral commanding the district, and by numerous naval officers and civilians. We learned that the *Discovery* and *Terra Nova* had arrived some time before, and had been paid off at Sheerness. The Admiral therefore issued instructions for the *Morning* to follow suit. Owing to the wretched state of our engines and boilers, the captain pointed out that this would be a difficult undertaking. ' But,' remarked the Admiral, ' you have your sails.' The captain explained that there were degrees of sailing vessels, and as the *Morning* sailed so much more sideways

210

UNFIT FOR THE CHANNEL

than ahead, he did not consider it prudent to attempt the passage under sail alone. Sailing in the open ocean was one thing, but in the narrow and crowded waters of the English Channel it was quite another matter, and the risk involved was too great. A hint thrown out for a cruiser to tow the *Morning* round was not taken up.

A party of engineers and artificers invaded the little ship next day, and after a thorough examination below, they were satisfied that our chief engineer's report was only too true. The boilers and engines were too far gone to justify incurring the expense of repairs. We were therefore towed over to Devonport to await instructions from the Admiralty, and moored amongst a formidable fleet of battleships and cruisers. Admittedly, the *Morning* was a small vessel, but the contrast with the great ships surrounding her was so striking that it was hard to realise that we had actually existed for over two years in such a diminutive-looking craft.

Evans was the happiest man in our company. His wife had arrived in England some weeks before, and was in Plymouth to greet him. It was a strange coincidence that the house they stayed at on the Terrace overlooking the Hoe was the one I lived in when I was eight years old.

The week we remained by the ship enabled us to bring all records and back work up to date. The 'spoils,' in the shape of cutlery, silver, books, &c., were gradually divided amongst the captain and officers, and the piano was unanimously voted to me. I sent it home in three parts, and rebuilt it later on. Although it was practically a wreck, I was proud to own an instrument that was associated with so many happy recollections; and certainly its history made of itself an interesting story.

At length word came from the Admiralty that the *Morning* was to be sold to a Dundee whaling firm. Two members of the Antarctic Committee from the Admiralty arrived at Plymouth, and, after paying off the crew, they took her over. That afternoon saw a busy and chaotic scene throughout the ship. We were packing up. It was a bewildering experience, and no one seemed

THE LAST OF THE 'MORNING'

to have enough bags and trunks to accommodate his personal belongings. The situation was much relieved, however, by the numerous bluejackets engaged discharging stores, &c., coming in for a generous supply of cast-off garments. Just before dusk, a dockyard tender came alongside, and was soon filled with a mountainous heap of luggage. We then looked round our little floating home for the last time. It was hard to tear oneself away! The entire crew embarked on the tender, and as she bore us away, we turned and gave three farewell cheers for the deserted vessel.

It was one of those sad occasions when that unaccountable lump would persistently stick in one's throat. We tried to be cheery, and even masked our feelings with indifference, but at heart we were very sad. The ship had been a home of innumerable happy memories and associations. She had served us faithfully, and had become endeared to us. What recollections crowded themselves into our minds as we looked back at the receding picture! The sun had set, and the grey October gloom was settling down fast. Overshadowed by the grim bulk

THE VOYAGES OF THE 'MORNING'

of Britain's pride, the brave little ship, with masts outstanding against the western sky, became fainter and ever fainter, until the enfolding night completely blotted her out from our lingering gaze. That was the last of the *Morning*, and we never saw her again. A few years afterwards she was lost in the Arctic, and her adventurous career was ended.

On the following morning there was an animated scene at the railway station, and, as the express steamed out for London, ringing cheers rang out from the homeward-bound crew. England, Morrison, and I remained two days longer to explain and hand over various documents and accounts.

This being satisfactorily concluded, we booked passage per express to London. England had so much gear to collect and pack that he did not think he could be ready to accompany us. The litter of articles in his room at the hotel certainly indicated this. Reluctantly we left him, surrounded by his innumerable belongings, to follow on, perhaps next day.

Morrison and I were seated in our 'smoker,'

A DESPERATE RUSH

and just as the starting whistle was about to blow, we were conscious of some commotion on the station platform. Out of curiosity we looked out of the carriage windows and were exceedingly astonished to see England with three porters wildly heaving boxes and bags into the luggage van. The train started, and, with a sprint equal to any professional hundred yards' pace, he hurled himself breathless into our carriage, weighed down and almost overpowered by a medley of bursting bags, skis, ski-stick, a penguin skin or two, and a rifle. Under his arm he gripped a home-made cabinet box in which his valuables were kept, and a cumbersome brass clock was suspended by a lanyard from his wrist. Altogether the effect produced was excruciatingly funny, and it took him nearly all the way to London to recover from his desperate rush.

Night had fallen when we reached the great city. It was fascinating driving through the brilliantly lighted, bustling thoroughfares once again. After our two years' absence, a period so crammed with unique experiences, the impression it made was not unlike the awakening from a wonderful dream.

THE VOYAGES OF THE 'MORNING'

At the station we parted. Morrison set off for Glasgow, England for Sheffield, and I for Blackheath. The captain was at Hull, the doctor at Edinburgh, and Evans and his wife in London. The temporary separation could hardly have been more complete.

It is unnecessary to attempt to draw a picture of the sweetness of home after our wanderings. A veil may, therefore, be drawn over a family party grouped round a cheerful fire, happily listening, far into the October night, to the story of the *Morning's* voyages as related in the foregoing pages.

CHAPTER XVIII

Expedition functions—Captain Scott's lecture at Royal Albert Hall—Graceful tribute to Captain Colbeck—Reception of Morning's crew at Hull—Royal Corinthian Yacht Club's dinner—Captain Colbeck's wedding—The sad farewell.

INTERESTING and enjoyable functions in connection with the Expedition were held during November and December. The officers from the three ships took up temporary quarters in London, and this resulted in many happy reunions.

Owing to the *Morning's* protracted voyage home, the authorities thoughtfully delayed the social proceedings until her officers could participate in them also.

Our first memorable meeting was the welcome-home dinner given by the Royal Geographical and Royal Societies at the South Kensington Hotel. There were numbers of eminent scientists present, and members of

THE VOYAGES OF THE 'MORNING'

the Geographical Societies from many parts of the Continent. Sir Clements Markham took the chair, and paid a glowing tribute to Captain Scott on the success of the Expedition. A feature of the meal was the last dish on the menu. This was a wonderfully devised facsimile of the *Discovery* made of ice-cream. It was carried round on a large tray by two waiters, and as the ship was gradually cut away, the masts tottered and fell among the wrecked remains.

Following the dinner was Captain Scott's maiden lecture at the Royal Albert Hall. The huge building was packed with a brilliant audience, and Captain Scott received a great ovation. A screen was hung across the huge organ at the back of the lecturer, on which the beautiful pictures were thrown from a lantern in the gallery opposite. For one not used to this sort of thing it was a trying ordeal, and Captain Scott was at first a little nervous. However, he soon warmed to the subject, and gave a splendid account of the three years' doings of the National Antarctic Expedition. By means of an electric button he communicated with the lantern operator when to change the

A LECTURING MISHAP

slides. At one part of the lecture he became enthusiastic on a description of penguins, and signalled for a penguin picture. Through some mistake, this did not appear, and as he proceeded—' Here you will notice these quaint birds in the act of—' He turned to the screen and was surprised to see a party of men harnessed to sledges. The audience laughed good-humouredly, but poor Captain Scott became very worried. To try and counteract the mistake, he immediately started to describe sledging work, but the operator, seeing his error, unfortunately switched on a group of penguins. Renewed applause. Captain Scott turned round to us, despair in his face, and remarked : ' The whole show is being messed up ! ' We cheerfully encouraged him, and assured him that he was doing excellently. He soon got over his embarrassment, and, fortunately, no more slips occurred. It was a great night for him, and he received not only the honour of obtaining the Royal Geographical Society's Gold Medal, but the Philadelphia Royal Geographical Society's Gold Medal was also presented to him by the American Ambassador.

THE VOYAGES OF THE 'MORNING'

A feature of particular delight to us was the presentation by Sir Clements Markham of a silver loving-cup to Captain Colbeck. It represented a globe on which were engraved the tracks of the *Morning's* voyages, the whole being supported by three inverted seals. It was a very graceful tribute to the relief ship's assistance in the enterprise.

As our commander and nearly all of the *Morning's* crew hailed from Yorkshire, the citizens of Hull arranged a reception in our honour at the Town Hall. The building was well filled, and many complimentary speeches were made. In the evening a dinner was given to the crew at the Sailors' Home, followed by a dance. The sailors had invited many of their friends, and these included several fishermen's wives. But a gloom hung over the gaiety. The day before, the Dogger Bank outrage had occurred, and I remember how concerned those poor anxious women were for their husbands' safety.

Perhaps the happiest function we attended was the annual dinner of the Royal Corinthian Yacht Club. This took place early in December at the Hotel Cecil. The *Morning*

THE ROYAL CORINTHIAN

was borne on the books of this club as a yacht, and similarly, the *Discovery* belonged nominally to the Royal Harwich Yacht Club. It was therefore appropriate that, at the annual dinner, the officers of the club's unique yacht should be included.

It was a magnificent banquet, with an attendance of possibly two hundred men. Prominent members of the Club were told off to look after us, and there was no hitch in this direction. We could not have been more hospitably treated.

The principal toast of the evening, after 'The King,' was 'The Captain and Officers of the *Morning*.' This was received with tremendous applause. The Commodore, in the chair, proposed the toast in the most complimentary and flattering expressions, and we fully appreciated the many nice things said. We felt that we had genuinely earned a certain amount of praise, and a little personal pride was, under the circumstances, quite excusable. But the ovation that accompanied the Commodore's toast was as nothing compared to the deafening uproar that greeted our fine commander, Captain Colbeck, as he

THE VOYAGES OF THE 'MORNING'

rose to respond. It was a wonderful reception, and we felt prouder than ever. It was with difficulty that he could commence, so persistent was the cheering. At last he was given a chance, and made a happy speech, skilfully worked up to a striking conclusion. He stated that he was proud to have commanded the largest yacht of the Club, and to have navigated her not only farther south than any other yacht, but farther south also than any other ship in the world. As the *Morning*, he continued, had sailed under the Royal Corinthian Yacht Club's ensign and burgee, and had flown them in the South Polar Regions, he thought that it was only fitting that these now historic flags should be returned and presented to the Club. As he led up to this last sentence, the actual flags referred to were slowly hoisted on two flag-poles lashed crosswise behind the Commodore's chair. The effect of this was electrifying, and the applause was deafening. Men rose to their feet, some stood on their chairs, to be better heard, perhaps. The accompanying waving of arms was distracting, glasses clinked, programmes were hurled into the air. Who

A WEDDING PARTY

said that Englishmen were cold and reserved ? They were absolutely carried away in their excited enthusiasm. The hoarse cheering gradually evolved itself into the usual musical honours, and culminated in three of the most rousing cheers I have ever heard. It was the greatest reception we could possibly have had, and my blood tingles still when I think of that memorable night.

About the middle of December we met for the last time. This was at our captain's wedding in Manchester. We foregathered at the same hotel on the eve of the marriage, and made merry until late into the night. But mixed feelings predominated. We knew that this was the end of our companionship.

The ceremony was, as usual, a combination of gravity and gaiety. While the organ was playing appropriate wedding music, and the bridegroom and best man were standing gravely awaiting the bride, the suspense was broken by the original Morrison. Our party were in a pew together, and he suddenly leaned over, and, nodding towards the captain, he remarked in a stage whisper: 'I'll bet he'd sooner be in a south-east gale!' The reception held at the hotel was gay, and after the customary speeches were over we charged our glasses once more. In a quiet corner of the room the captain and officers of the *Morning* then impressively bade each other farewell, and in silence drank for the last time together.

THE END.

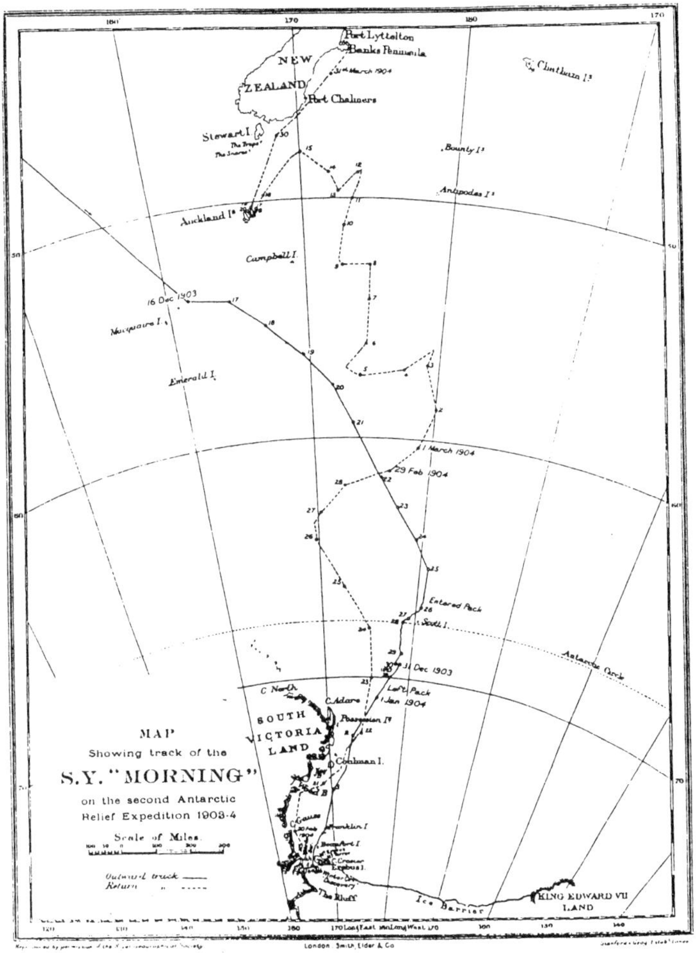
MAP
Showing track of the
S.Y. "MORNING"
on the second Antarctic
Relief Expedition 1903-4
Scale of Miles
Outward track
Return
NEW ZEALAND
Port Lyttelton
Banks Peninsula
Port Chalmers
31st March 1904
Chatham Is
Stewart I.
The Snares
Bounty Is
Antipodes Is
Auckland Is
Campbell I.
16 Dec 1903
Macquarie I.
Emerald I.
1 March 1904
29 Feb 1904
Entered Pack
Scott I.
31 Dec 1903
Antarctic Circle
Left Pack
1 Jan 1904
C North
C. Adare
SOUTH VICTORIA LAND
Possession Is
Coulman I.
C. Gauss
Franklin I
Beaufort I
C. Crozier
Erebus I.
Discovery
The Bluff
Ice Barrier
KING EDWARD VII LAND
170 Long. East 180 Long. West 170
London: Smith, Elder & Co